The Great Inventions

by Ralph Stein

The Great Inventions

A Ridge Press Book / Playboy Press

Editor-in-chief: Jerry Mason
Editor: Adolph Suehsdorf
Art Director: Albert Squillace
Associate Editor: Ronne Peltzman
Associate Editor: Joan Fisher
Art Associate: David Namias
Art Production: Doris Mullane
Picture Research: Marion Geisinger

Published simultaneously in the United States and Canada
by Playboy Press, Chicago, Illinois. Printed and bound in Italy
by Mondadori Editore, Verona.

Library of Congress Cataloging in Publication Data

Stein, Ralph
The great inventions.

Includes index.
1. Inventions—History. 2. Inventors—Biography.
I. Title.
T19.S73 609'.034 76-12484
ISBN 0-87223-444-4

Picture Credits

Abbreviations

ILN *The Illustrated London News*
LC Library of Congress
NMM National Maritime Museum, Greenwich, England
NYHS New-York Historical Society
NYPL New York Public Library Picture Collection
RS Collection of Ralph Stein
SI Smithsonian Institution, Washington, D.C.
SM Science Museum, London
SML Photographed by Ralph Stein at Science Museum Library, London
SSM Singer Sewing Machine Company
TAE The Thomas A. Edison Foundation Museum, West Orange, N.J.
WVD Collection of Willis Van Devanter
Yale Yale University Library, New Haven, Conn.

Chapter 1
LC: 13. Mercedes-Benz: 16 (ctr). NYPL: 15 (bot rt), 16 (bot), 17. RCA: 14 (bot). RS: 14 (top), 15 (ctr), 16 (top lt). SM: 15 (top, bot lt). SSM: 16 (top rt).

Chapter 2
Chessie System, Baltimore: 76 (top), 79. Detroit Institute of the Arts: 48. ILN: 18-19, 67 (photograph by Richard Cooke). Museum of Science and Industry, Chicago: 75 (bot). NMM: 44 (mid rt). NYPL: 49 (Manuscript & Archives Division), 53 (top: I. N. Phelps Stokes Collection of American Historical Prints), 53 (bot), 56, 61 (lt), 80. NYHS: 50-51, 52. Railroad Museum of Pennsylvania, Strasburg: 78. RS: 24. SI: 42-43, 44 (bot), 46, 47 (top), 54 (top lt). SM: 21, 22, 23, 26, 28-29, 31, 32 (bot), 33, 34, 35, 36, 38, 39, 40, 41 (top), 44 (top, mid lt), 54 (top rt, bot), 55, 58-59, 60, 62-63, 64, 65, 66, 68, 72 (lt), 73, 75 (top). SML: 70, 71. Yale: 30, 32 (top), 41 (bot), 47 (bot), 61 (rt), 72 (rt), 76 (ctr, bot).

Chapter 3
A. T. & T. Company Photo Service: 104 (top). Insurance Company of North America: 82-83. LC: 100 (top), 102, 107, 114, 115. Marconi Company Ltd.: 118 (lt), 119, 120 (top), 122, 123, 125. NYHS: 110. RCA: 128, 133 (top), 134 (top), 135, 136, 137. SI: 86 (bot), 94, 98, 113, 116, 126, 129, 133 (bot), 134 (bot). SM: 85 (mid lt, bot), 86 (top, mid), 87, 88 (lt), 90 (bot), 91 (bot), 92, 93 (top, bot rt), 96, 104 (bot), 105, 106, 111, 117, 118 (rt), 120 (bot), 124, 130, 133 (ctr). Syracuse University Art Collection: 100 (bot). TAE: 127. Yale: 85 (top, mid rt), 88 (rt), 90 (top), 91 (top), 93 (bot lt), 97, 99, 101, 103, 108.

Chapter 4
NYPL: 147. RS: 153 (rt), 154, 155 (lt), 156. SM: 140, 141, 145, 148, 149 (rt), 150, 152, 153 (lt), 155 (rt), 157. WVD: 138-39, 142, 143, 144 (rt), 146. Yale: 144 (lt), 149 (lt), 151.

Chapter 5
RS: 162 (bot). SI: 180, 182, 183, 186 (lt), 188, 189. SM: 160, 161, 162 (top), 163, 164, 166, 167, 168 (bot), 170, 171, 172, 173, 174, 175 (bot), 176, 177, 178, 179, 181, 184, 185, 186-87. SML: 158-59. Yale: 168 (top), 169, 175 (top).

Chapter 6
The Alfred Stieglitz Collection, Art Institute of Chicago: 206 (bot), 207 (lt). George Eastman House, Rochester: 195 (lt, ctr), 196 (lt, rt), 207 (rt), 213 (top), 217 (bot). NYPL: 201, 218 (top), 220 (bot). Polaroid Land Camera Corp.: 213 (ctr, bot). RS: 190-91, 198 (bot), 211 (top rt, bot), 213 (bot lt), 215. SM: 192, 193, 194, 196 (ctr), 197, 198 (top), 199, 200, 202, 203, 204, 205, 206 (top), 208, 209, 210, 211 (top lt), 212, 214, 216, 217 (top lt), 218 (bot), 219, 220 (top, ctr), 221 (lt). Société Française de Photographie, Paris: 195 (rt). Stanford University Museum of Art, Muybridge Collection: 217 (top rt). Yale: 221 (rt).

Chapter 7
Brown Brothers: 230 (bot), 239. The Metropolitan Museum of Art, gift of I. N. Phelps Stokes, Edward S. Hawes, Alice Hawes, Marion Augusta Hawes, 1937: 224. NYPL: 228 (bot lt), 231, 238. RS: 236 (no. 2), 242 (bot). SI: 235 (nos. 5, 6), 243, 244, 246 (top). SM: 225, 226, 227, 228 (except bot lt), 229, 232, 233 (lt), 234, 235 (nos. 3, 4, 7), 236 (nos. 3-6), 237, 240, 241 (rt), 242 (nos. 1, 2), 247 (nos. 7, 8), 248. SSM: 230 (top, mid). TAE: 222-23, 246 (bot). U.S. Patent Office: 236 (no. 1). Yale: 233 (rt), 241 (lt), 242 (nos. 3, 4), 245, 247 (nos. 3-6).

For Muriel

Contents

5 We Take to the Skies 160

6 Light Does the Drawing 192

7 The Little Servants 224

8 Epilogue 249

Acknowledgments

During trips to London I have seldom failed to visit the Science Museum in Exhibition Road, South Kensington. One of the world's great museums, it is the repository not merely of models and reproductions of the great inventions, but of the actual devices themselves. There I have examined Stephenson's locomotive *Rocket,* Fox Talbot's cameras, Hooke's seventeenth-century microscope, Cooke and Wheatstone's telegraphic instruments, Alcock and Brown's transatlantic Vickers-Vimy biplane of 1919, and hundreds of other examples of mankind's creative imagination.

On a recent journey to the museum, I was not only enjoying myself, I was working on this book and delving into the museum's remarkably complete photographic archives, many of whose prints have been struck from glass negatives dating back to the earliest days of photography.

Among the many people on the Science Museum's staff who helped me, I owe special thanks to Mr. Walter Winton, Keeper of Engineering Collections; Lt. Cdr. W. J. Tuck, Deputy Keeper, Aeronautics Collections; and Mr. P. R. Mann and Mr. R. E. Burrell of the Aeronautics Collections. Miss Annette Crane of the museum's Information Department gave me a tremendous amount of help in finding the many photographs I needed.

Other organizations in England were also most cooperative: the National Maritime Museum in Greenwich, the National Gallery in London, and the Marconi Company. I owe a particular debt to Mr. Peter ffrench-Hodges and Mr. Edmund Antrobus of the British Tourist Authority for their great assistance. Miss Adrianne Le Man, art editor of the *Illustrated London News,* most graciously allowed me to use photographs from her files.

In this country I naturally leaned heavily on the Smithsonian Institution for assistance. Mr. Silvio A. Bedini, deputy director of the National Museum of History and Technology, and Miss Carol Forsyth of the Photographic Department were generous with their help and knowledge.

Much of my research was done in Yale University's Sterling Memorial Library. I am grateful to Mr. Harry Harrison, head librarian of the Circulation Department, who devoted much time and effort to helping me find the information I needed. Mr. Dale Roylance, Curator of the Arts of the Book Department, and Mr. Willis Van Devanter of Rectortown, Virginia, made it possible for me to photograph part of Mr. Van Devanter's superb collection of prints devoted to the history of the bicycle. To them, also, my sincere thanks.

R. S.

Tumbrils' End
Westbrook, Connecticut

NEW YORK
VIEWS, ARCHITECTURE, &c
STATUARY, STILL LIFE &c
SINGLE PORTRAITS & GROUPS
Mr. BEARD,
(SOLE PATENTEE)
OF THE
DAGUERREOTYPE
OR
Photographic Portraiture.
34, PARLIAMENT St. WESTMINSTER,
85, KING WILLIAM St. CITY,
&
ROYAL POLYTECHNIC INSTITUTION
309, REGENT ST. LONDON.
Here is an agent in Photography by which the much prized family Portrait the old cherished Picture may be multiplied at a trivial cost & with such exactness & truth as to establish an identity of character the minutest detail may be perfectly represented.
As well for momentary expression of countenance as for drapery in elaborate patterns, for chased Armour, carved furniture, articles of Vertu whether of Marble or Silver, indeed for every description of still life the Photographic principle has been successfully employed.
It was color that was wanting to crown the improvements, Then was obtained what had been so often sought for a speaking likeness. No matter how delicate the tint, how bright the glow, the color was communicated.
JONATHAN HOLLS'S STEAM-BOAT

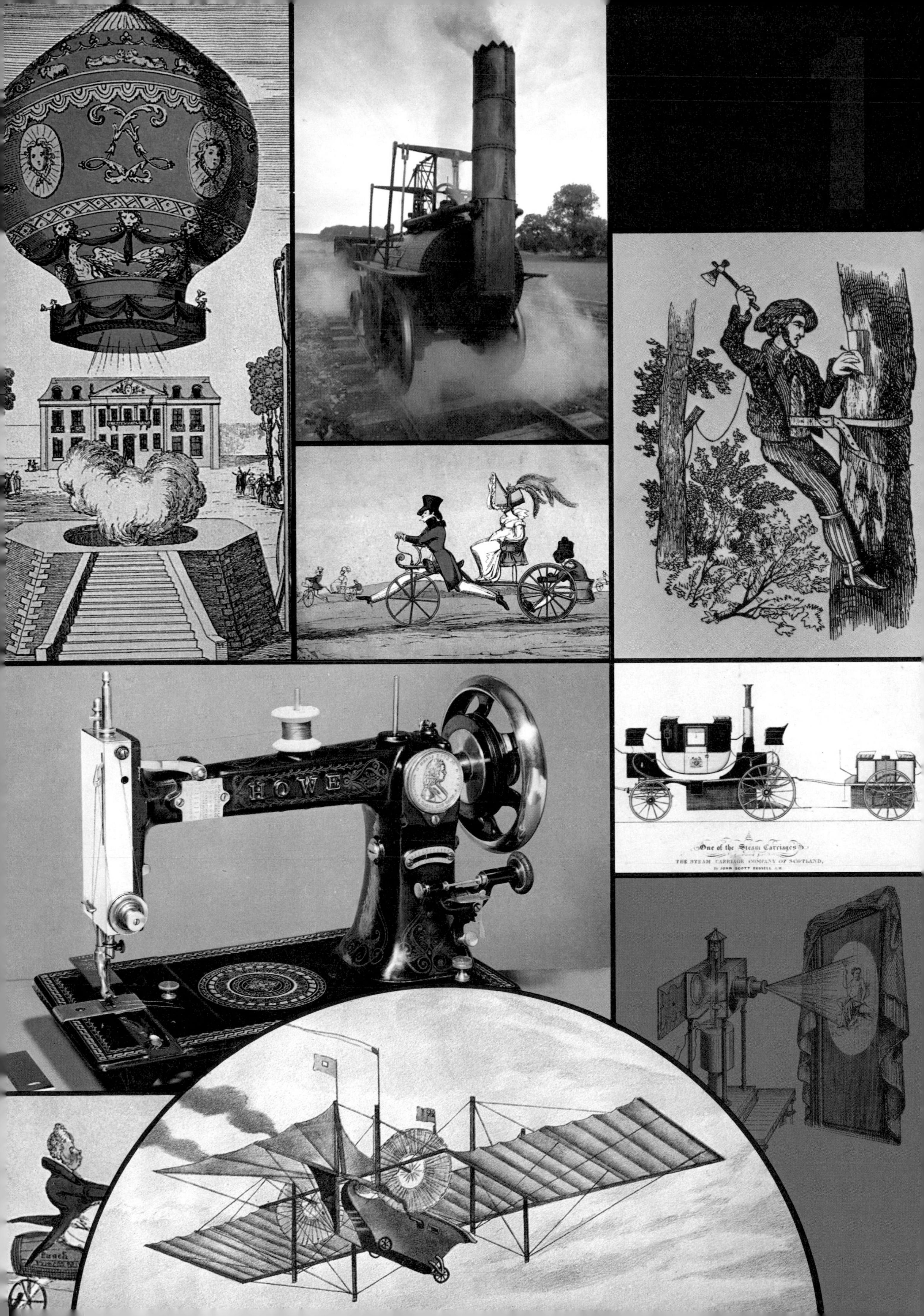
HOWE
One of the Steam Carriages
THE STEAM CARRIAGE COMPANY OF SCOTLAND,
JOHN SCOTT RUSSELL

1 Introduction

Man's journey through the ages has been eased, accelerated, and frequently complicated by his unique and irrepressible knack for invention. From the stone ax and clay pot of prehistory to the electron microscope, computer, and spacecraft of the twentieth century, the creations of his innovative intelligence have been ingenious, phenomenal, and occasionally —for good or ill—of world-shaking significance.

This book is concerned with inventions as they have sprung into being in the most recent few minutes of that journey. For the source of many children of man's imagination is now beyond recall. Much as historians may wonder about the geniuses who invented the wheel, the hammer and nail, and written language, or discovered how to make wine, weave cloth, cut diamonds, or navigate by the stars, it is not likely that anything will ever be learned about them.

The past two hundred years, however, are a different matter. Here is where many of the basic components of contemporary life were conceived and developed. Here is where the pattern and style of our existence was begun. And how it all happened is well documented.

With so much to choose from, we decided to concern ourselves essentially with those great inventions responsible for changing the slow, agrarian eighteenth-century world into the frantic, mechanized world of the twentieth. Further, we have been choosy about which of the many inventions we would deal with. Only those basic notions that radically affected the way people lived, and still live, have been included. And, except for a short discourse on Robert Fulton's submarine experiments, we have avoided weapons of war.

Most of the inventions discussed here were the fruits of that amazingly vital period between the latter years of the eighteenth century, when the Industrial Revolution began, and the start of World War I. It might even be said that almost every important invention came during that great century— the nineteenth—if we stretch it a few years at each end.

The Industrial Revolution was, at first, British. By the early eighteenth century England was the world's leading industrial country. But its manufacturing was not done in factories. Cottagers spun cotton and worked handlooms in their own smoky little houses, or in the houses of village contractors. Other articles were made in the small, private shops of craftsmen—carpenters, blacksmiths, potters, wheelwrights, and the like.

By mid-century change was in the air. A growing colonial empire was clamoring for goods. And there was money around for capital investment in new ways of making things and for new means of transportation and communication. But technological progress could be made only as fast as scientific knowledge advanced and mechanical skills were refined.

The seventeenth century had produced brilliant discoveries in physics, chemistry, and mathematics, but theory was generally in advance of fact. The eighteenth century undertook to consolidate this knowledge, although professional scientists were few and all elements of the scientific method had to be created. Standards of measurement had to be defined: weights, dimensions, and most particularly time. Laboratory techniques had to be devised. Scientific equipment had to be designed and made, and scientific information had to be disseminated. Late in the century, engineers began to be able to utilize the scientific principles formulated in the seventeenth century, and a group of younger scientists, in places like the universities of Glasgow and Edinburgh, began cooperating with engineers in solving technical problems.

Learned groups, like the Lunar Society and the Royal Society in England and similar societies in Europe, also furthered the advance of technology by forum discussions of scientific matters. Scientists, inventors, and engineers were there able to put forward new ideas to the bankers and merchants who had the capital to invest in new projects. And bodies like the Royal Institution of Great Britain and the Conservatoire des Arts et Métiers in Paris created popular excitement over new technological and scientific advances through public demonstrations and lectures. Dilettantes of the middle and upper classes flocked to the lectures, and it became fashionable to pursue scientific research, especially electrical experiments, in elegant English country houses and French châteaux.

But real progress took time. When James Watt built his first steam engines he nearly went out of his mind trying to

Preceding pages: Almost every device we use today, from the sewing machine to the telephone, was first thought of by some bright fellow who lived during the years between the American Revolution and the First World War. Below: Old lithograph depicts nineteenth-century progress.

Only the windmill on the opposite page dates from before the Industrial Revolution. The locomotive, the phonograph, the electrical machine, and the telephone came after. The artist's imaginary traffic jam of balloons has not, happily, come to pass.

find workmen who could make cylinders and pistons that fit each other and were steam-tight. Even such small parts as bolts and nuts were at first wildly inaccurate—no two were of exactly the same size and pitch. Only after Henry Maudsley invented the screw-cutting lathe in 1797 did the accurate machining of interchangeable parts begin to be practical. (A year later Eli Whitney developed the production of firearms using interchangeable parts at a factory near New Haven, Connecticut.)

The first changes in Britain's old handicraft system were in the manufacture of textiles. Richard Arkwright's cotton-yarn spinning machine and James Hargreaves' spinning jenny were the starting points of a long line of machines that mechanized the textile industry, taking it out of the cottage and into the factory, and taking the workers out of their villages and into industrial towns.

Increased manufactures resulting from new inventions brought greater prosperity and rapid population growth. Bigger populations needed better roads, better road transport—better communications. This continuing expansion of needs and their satisfaction still goes on. The Industrial Revolution may have slowed down, but it is not over.

"Invention," someone has said, "is the mother of necessity." Each new invention not only depends on the inventions and techniques that come before it, it also spawns still newer inventions. The steam engine, for example, brought the steamship and the railway. The need for efficient control of train movements led to the development of telegraphy. Looking for improvements in telegraphy, Bell happened upon the telephone.

PARIS · BERLIN · ST PETERSBOURG
117
2627

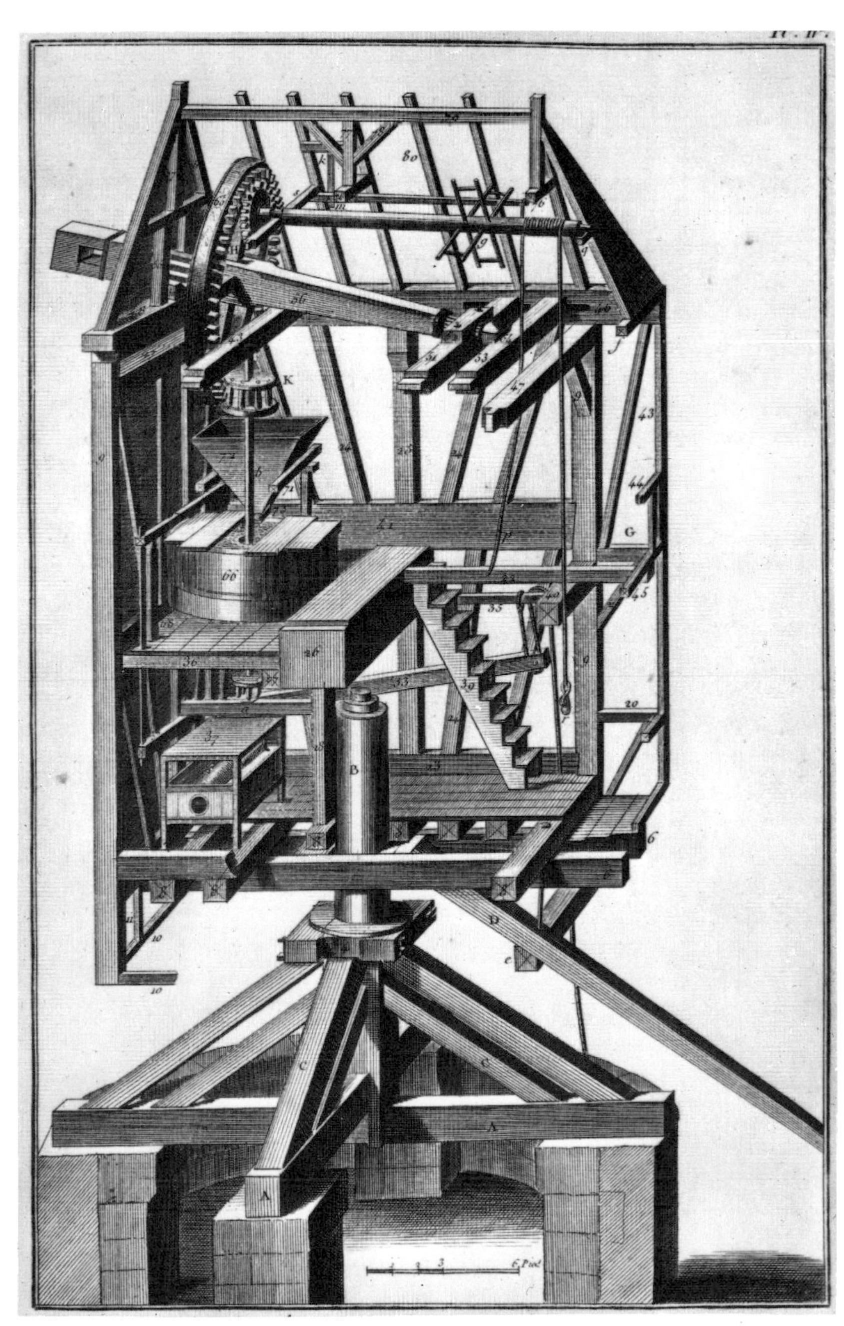

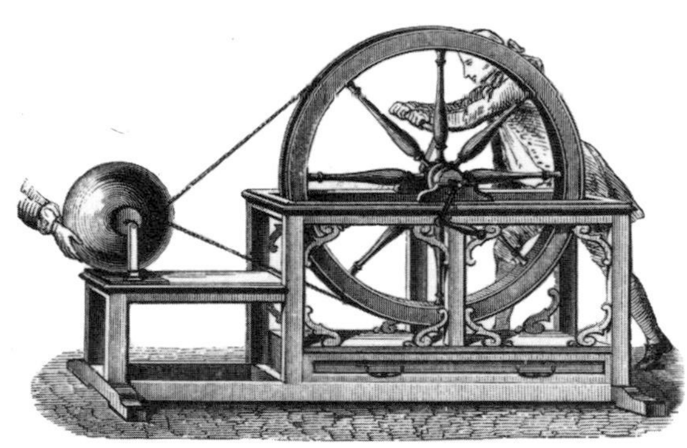

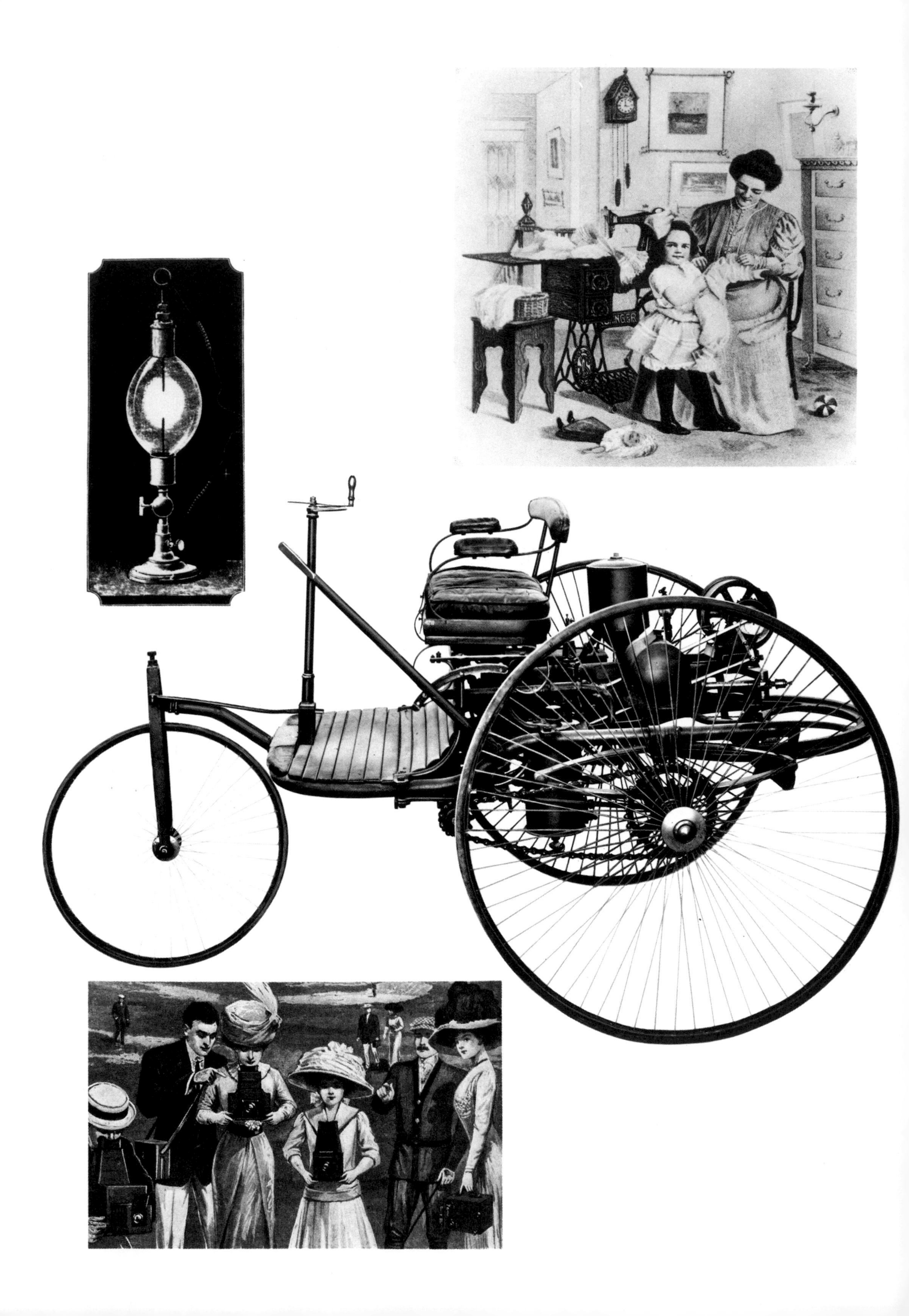

The electric arc light, the family sewing machine, the motor vehicle, the amateurs' cameras (these are Graflexes, about 1910), and the typewriter all came from the fertile brains of eager nineteenth-century experimenters.

Inventions are not made in a vacuum. The world goes on while the inventors experiment. What happens in the laboratory is to a great extent determined by what goes on outside, and vice versa.

James Watt's steam engine came as Britain was losing her American empire and gaining India. Steam would run the looms of Manchester which wove the cotton cloth exported to India. And India paid for the cloth with the gold that helped fuel Victorian prosperity.

In America the newfangled steamboats started churning their paddle wheels up the Ohio and Mississippi rivers just as the young United States was opening up the new lands west of the Alleghenies. By the 1840s, as the country was expanding westward to the Mississippi, the new railroads came just in time to bring in settlers and to link the new midwestern towns with the Atlantic seaboard.

Later nineteenth-century inventions—the telegraph, the sewing machine, the telephone, the automobile, photography, the typewriter, and the rest of the tools of our twentieth-century lives—came during a comparatively peaceful time, broken only by the Crimean War, the American War Between the States, and the Franco-Prussian War, conflicts that made use of recent inventions adapted for war—the telegraph, the balloon, and photography.

During the nineteenth century, the Industrial Revolution spread to almost every European country. Germany, in particular, became the most industrialized country on the continent. Her efficiently mass-produced exports seriously rivaled those of Britain and the United States. The resulting violent competition for world markets was one cause of the First World War.

And that war in turn used the recently invented aeroplane, wireless, and internal-combustion engine for its bloody purposes.

Why was the nineteenth century the age of invention? The answer is obvious: it was during that century that the materials, tools, and techniques inventors needed to carry out their ideas became increasingly available.

But there was another reason. The nineteenth century was a time of confidence and comparative peace. The world, people were convinced, was getting better and better. Every week *Scientific American* gave evidence of that improvement by bringing news of marvelous new devices. New ideas were stimulated, people were spurred to new efforts—there was an *atmosphere* of inventing. Inventors (the successful ones) were the heroes of the day. Edison was perhaps the biggest hero of them all. The press hung upon his every word. And mechanics, artists, farmers, bicycle repairmen not only wanted to be like the heroes they read about, they wanted to make money like them, too.

In schoolbooks perpetuating the not entirely true stories about Robert Fulton inventing the steamboat, Edison inventing movies, the phonograph, and the electric light, and Watt inventing the steam engine, inventors were made out to be noble fellows bravely surmounting all odds. They all seemed to have the high ideals of Eagle Scouts.

In fact, they were more human than that. Though there were, of course, exceptions—Robert Fulton, for example, seems to have been an ingratiating smoothie with a fine talent for gaining the confidence of the big wheels of his time, and Alexander Graham Bell was as nice a man as could be—the inventors seem in general to have had peculiar personalities. They were more cantankerous, stubborn, and unlovable than most people.

One cannot blame them for not being sunny fellows. They lived harried, unhappy lives while doing their inventing. Morse of the telegraph and Howe of the sewing machine almost starved to death while promoting their inventions. Yet, later, when they had made their bundles, they didn't turn out to be nice chaps at all. Morse was a right-wing "know-nothing," a copperhead who tried to have Lincoln turned out of office. Howe became a pompously, overbearingly rich man.

Why did these fellows become inventors in the first place? What makes people invent?

Some people just can't help themselves. They seem to *have* to invent. James Watt, successful and revered and in no need of money after perfecting the steam engine, devoted his last years to trying to invent a sculpture-copying machine.

But there was, and still is, one overriding reason why people invent. They hope to get rich.

2

2 New Muscle for the World

The World Before Steam

The Industrial Revolution and its driving force—steam—wrought great changes in what was for most people a time of heavy, man-killing labor, dirt, disease, and despair.

The myth of country people living in bucolic bliss dies hard. But agricultural workers of the eighteenth century were little better than slaves who worked His Lordship's land as tenants from daybreak to darkness. They were hardly better off than the serfs of 1066—or for that matter the slaves of the Pharaohs. The townsman, the artisan who made shoes or clothing or pots and pans, was only a little better off. Sewage ran down the middle of his mean cobbled streets. At night he might have a candle to push back the darkness, a luxury the farm laborer could not afford. Only the very thin upper crust, the nobility and landed gentry, lived what might be considered tolerable lives.

For rich and poor alike in the world before 1800 there were few sources of power, few machines, and few manufactured products of any kind. The prime sources of power were the muscles of man and beast, and for fixed locations the windmill and waterwheel. The basic machines—mostly operated by hand or by a treadle—were lathes, presses, pumps, cranes, and looms. Many items, of course, were handcrafted by artisans, but manufacturing processes were applied only to products like textiles, books, firearms, and pottery.

There were, in the eighteenth century, intricate and beautifully constructed mechanisms, but few people used or even saw them. Microscopes, for example, had achieved a high degree of precision by the 1750s. George III owned one of the more complex and intricate of them. But King George was a patron of the sciences and had a large collection of "philosophical instruments," many of which can be seen today in London's Science Museum. It was during the eighteenth century, too, that John Harrison constructed his series of unbelievably complicated timepieces for finding longitude at sea.

The French, although they didn't go through an industrial revolution until after their political revolution, were no slouches either when it came to intricate devices for the very few. Their clocks and watches were second to none. And before 1800 Joseph Marie Jacquard built incredibly intricate looms for weaving silk into highly detailed pictorial patterns. The looms were programed by punch cards controlling four hundred strands of warp. Few devices anywhere, however, equaled the automata built by Jacquet-Droz *père et fils* and the *mécanicien* Kintzing for the edification of Marie Antoinette and her rich and royal friends. These remarkable toys, which took years to construct, wrote, played the *tympanon,* and indulged in other less artistic activities with near-human ease and smoothness of action.

Yet by mid-century a fresh breeze began to blow through the western world. The Age of Reason began to make its way. In France, social and political revolution was the result. In America it was revolution and independence from Britain. In England it was the "Industrial Revolution."

It was in the textile industry that this beginning of a mechanical civilization started. Richard Arkwright's yarn-spinning frame of 1767 was among the first machines which changed England from a country of cottage industries to one of factories. Cottage industries could not supply the growing demand for goods. And the hunger for manufactured goods came not only from England's growing population in a time of comparative peace, but also from her new colonial empire stretching from Massachusetts to Madras.

Textiles were one thing. Iron was even more important and marvels were accomplished with it, although cast iron was too brittle to stand much stress and wrought iron too soft. Both would be supplanted by steel—tough, precise, unyielding—which improved the efficiency of almost everything made of metal. But in the world before steam, steel was difficult and expensive to smelt and was made only in small batches for such specialty items as cutting tools and machine parts. Iron, with all its imperfections, was the best metal available.

At first charcoal was used in its smelting. But by the beginning of the eighteenth century the forests of England

Preceding pages: Opening day, September, 1825, of the world's first public steam railway, the Stockton & Darlington. Left: John Harrison's first marine chronometer. Below: Portrait of Joseph Marie Jacquard, woven on one of his punch-card programed looms.

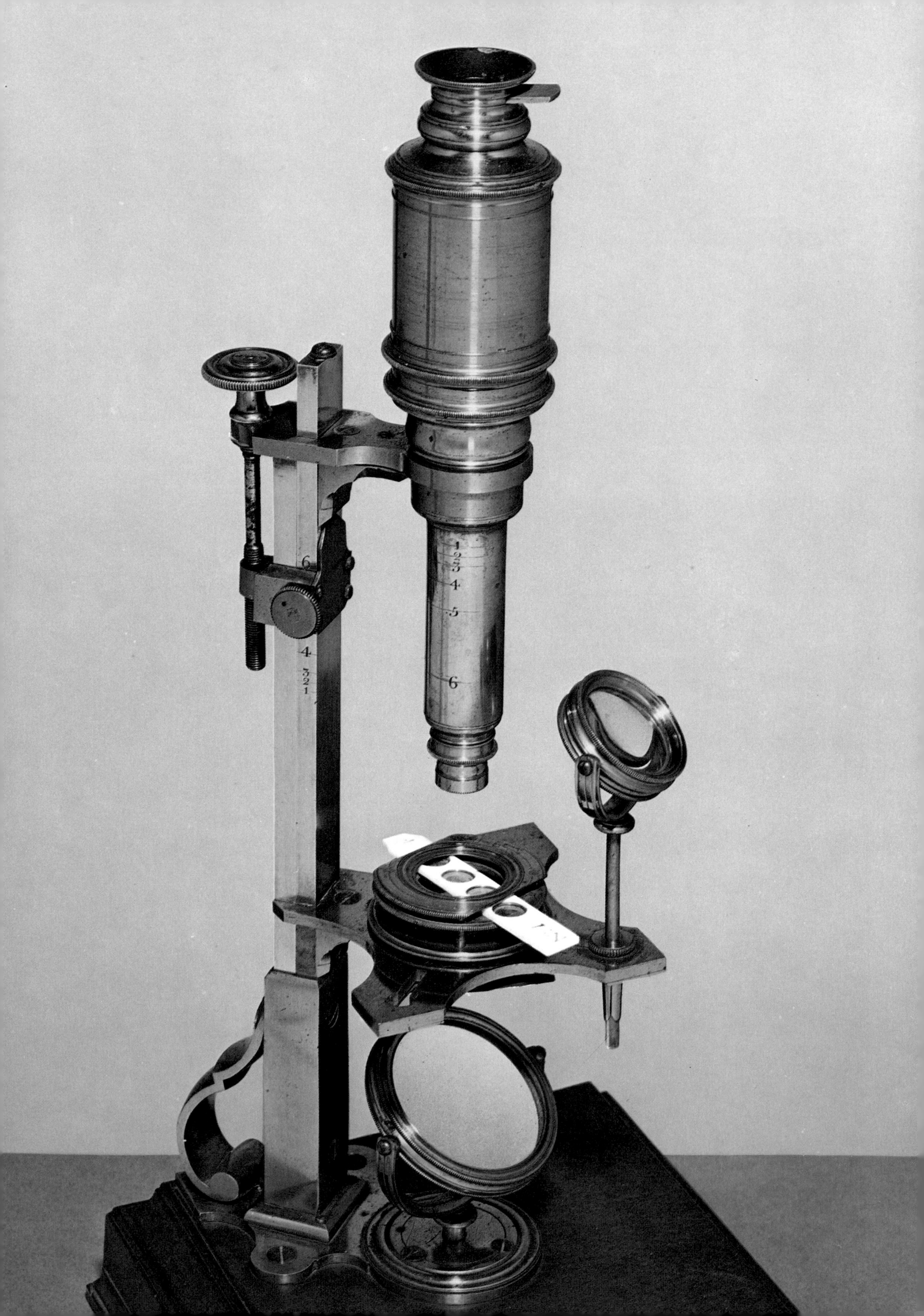

Opposite: The 1744 Cuff microscope exemplified the high degree of precision attained by eighteenth-century instrument makers.
Below left: Coach travel was painfully slow in 1706.
Below right: Richard Arkwright's yarn-spinning machine of 1767.

YORK Four Days
Stage-Coach.

Begins on Friday the 12th of April 1706.

ALL that are desirous to pass from *London* to *York*, or from *York* to *London*, or any other Place on that Road; Let them Repair to the *Black Swan* in *Holbourn* in *London*, and to the *Black Swan* in *Coney-street* in *York*.

At both which Places, they may be received in a Stage Coach every *Monday*, *Wednesday* and *Friday*, which performs the whole Journey in Four Days, (*if God permits.*) And sets forth at Five in the Morning.

And returns from *York* to *Stamford* in two days, and from *Stamford* by *Huntington* to *London* in two days more. And the like Stages on their return.

Allowing each Passenger 14l. weight, and all above 3*d.* a Pound.

Performed By
Benjamin Kingman,
Henry Harrison,
Walter Baynes,

were almost gone. Coal had to be the new fuel. And great ironmasters showed the way to use it.

But coal had to be transported. At first water transport—canals—was the answer. Water, too, had powered the textile mills. James Brindley was the great canal builder. He and others built hundreds of miles of them connecting every major town in a huge spider web of water that sometimes crossed natural rivers on amazingly high aqueducts. The canals carried not only coal but dishes from Tom Wedgwood's potteries, bales of cloth, rolls of tobacco, and kegs of rum. The Industrial Revolution started to hum.

But it didn't start to roar until the steam engine came along. As we shall see, the steam engine first had to pump out the perennially flooded shafts of coal mines. And then it had to take over from the water-driven mill wheel, mostly because it was more reliable. No longer could drought or ice shut a factory down.

When steam was applied to trains on rails and ships on the seas, as well as to the burgeoning factories, England at last sped ahead into the unparalleled prosperity of the Victorian Age—prosperity mostly for the upper classes and for a newly risen middle class. The poor went into the

grim factories, steel mills, and mines, where they toiled, little better off than they had been as tenant farmers.

The new railroads caused the cities to expand almost explosively. London, for example, had a population of only some 600,000 in 1750, since it was impossible for wagons to bring in enough food for a greater number of mouths. By 1850 the population had quadrupled to more than 2,500,000. Railways fed the millions.

Steam power, steam railways, steamships spread the industrial revolution to almost every other western country during the nineteenth century. We would still be living in the Middle Ages if it were not for Newcomen, Watt, Trevithick, Stephenson, and their farseeing colleagues.

The Steam Engine

In 1777, soon after Matthew Boulton became James Watt's partner in building the steam engine Watt had invented, he was approached at a royal levee by George III.

"Ha! Boulton," said the king. "It is long since we have seen you at court. Pray, what business are you now engaged in?"

"I am engaged, your Majesty, in the production of a commodity which is the desire of kings."

"And what is that? What is that?"

"POWER, your Majesty!"

Until further enlightened, good old George may have thought that Boulton was talking about the kind of power needed to deal with those recalcitrant colonials across the Atlantic—the power of redcoats and Hessians. But Boulton had a far more important kind of power in mind—a power which would change the world.

The steam engine was more important than any invention before or since. It ended man's dependence on human and animal muscle power and freed him from the vagaries of wind and water power. And, above all, unlike windmills and water wheels, the steam engine was portable. It could be put to work almost anywhere.

Although the steam engine is some two hundred years old now, it still does the world's heavy work. It runs electric-power plants—whether fed by oil, coal, or atomic fission. It powers ocean liners and nuclear submarines, and quirky inventors are again trying to harness it to automobiles.

Certainly James Watt is the father of the reciprocating steam engine, but people at least as far back as the ancient Greeks fiddled around with steam.

Among the many steam-powered gadgets used by priests and royal flunkies to impress the ignorant and credulous populaces of the ancient world, the best known was that mentioned in a book by Hero of Alexandria, who lived a century or so before Christ. This was the aeolipile, or Ball of Aeolus (the god of winds). It was a hollow metal sphere mounted on a vertical axle with a pair of suitably angled hollow tubes projecting from it. The sphere was partly filled with water and a fire lighted under it. Steam spurting from the tubes caused the sphere to spin merrily. (In the seventeenth century Isaac Newton proposed a car propelled by an engine using such a reaction, thereby demonstrating his First Law of Motion: "Every action has a reaction.")

In the sixteenth century a translation of Hero's book appeared in Bologna and excited new interest in steam power, but none of the experimenters of the Renaissance succeeded in harnessing the spinning aeolipile. However, an Italian architect named Branca did leave us a plan for using a jet of steam to spin a turbine he proposed to gear to what looks like a garlic-pounding machine.

Other men realized that steam, after filling a sealed vessel, would create a vacuum when condensed. A cork in a flask filled with boiling water and steam would be pulled in when the steam cooled. Such a vacuum could be put to work pumping water, for example.

Denis Papin, a French Protestant who left his native country to work in England, Italy, and Germany in order to avoid religious persecution, seems to have been the first man (in about 1690) to have experimented with moving a piston in a cylinder by means of the expansive power of steam. He tried

Opposite: Fanciful wood engraving of the steam turbine planned by a sixteenth-century Italian architect named Branca. Overleaf: Denis Papin was the first man to use steam to move a piston within a cylinder.

gunpowder first, and luckily he escaped the consequences of such a rash experiment. Papin's vertical cylinder was also his steam boiler. He injected water into the bottom of the cylinder and lit a fire under it. When the water boiled, the steam pushed the piston and the piston rod up. The rod was then held in place by a catch. He then removed the fire, causing the steam to cool and condense, thus forming a vacuum under the piston. Releasing the catch caused the piston to be sucked downward with a force powerful enough to do useful work. A rope attached to the piston rod could, via pulleys, lift weights. If the piston rod were coupled to one end of an oscillating beam with a pump at its other end, it could pump water. Papin's first model had a cylinder with a 2½-inch bore. It raised a sixty-pound weight once per minute. Papin was on the right track but his methods were impractical, to say the least. Imagine having to remove the fire between each stroke of an engine! Furthermore, poor Papin had to depend on the crude workmanship of his day. It would be a hundred years or more before machinists had tools of sufficient accuracy—and competence to use them—to assure a decent fit between a piston and a cylinder. But Papin's most serious error was his attempt to use his cylinder as a boiler, too.

Thomas Savery, an Englishman born in 1650, didn't make that mistake. He used a separate boiler for his "fire engine." Savery, a Cornishman, was familiar with the terrible problem Cornish miners had with water in their workings. Hand pumps, horse-powered chains of buckets, water wheels, windmills were of no avail. Savery's solution was his "Invention for Raising Water and Occasioning Motion to all sorts of Mill Work by the Impellent Force of Fire which will be of great use and Advantage for *Draining* Mines, Serveing Towns with Water, and for the working of all Sorts of Mills where they have not the Benefit of Water nor constant Windes."

Savery's "fire engine" had no moving parts at all. It was a collection of tanks and boilers connected with a cat's cradle of pipes and valves.

In essence, he used two tanks—one a boiler to make steam, the other for condensing the steam by cooling it with cold water. His condenser had a stopcock and a pipe leading to the water that was to be raised. After he condensed his steam, thereby producing a vacuum, he opened the stopcock to the atmospheric pressure above the water in the mine, and the water was sucked up about thirty feet into the condenser. He then forced fresh steam into the condenser, and the pressure shot the water out of the condenser and up a pipe to the height required. By recondensing the steam now in the condenser, the cycle was repeated. By opening and shutting the multifarious valves, a diligent operator could, Savery claimed, make the cycle repeat itself about four times a minute.

Savery's "fire engine" failed as a pumper-out of mines, for in practice it could lift water only about sixty feet and some mines were three hundred feet deep. Further, Savery suffered from bad technology. His valves fitted poorly, the joints of his pipes leaked, solder melted, and the amount of fuel used was fantastic—more than twenty times as much as would be used today for the same amount of work.

Still, his engines had a certain success. In 1712 at least one was used at Campden House in Kensington, London, for supplying a house with water pumped from a well.

The first real steam engine, practical enough to perform work commercially, was Thomas Newcomen's "atmospheric" engine. Newcomen was a Devonshire man, born in 1663 in Dartmouth, and an ironmonger for whom the operators of the Cornish tin and copper mines were important customers. Their continuing problem and complaint was water in the mines. Savery's engine hadn't worked for them, nor had the panaceas of many other inventors. In Newcomen's day they were still fighting the water with horse-operated chains of buckets and the water was still winning.

Newcomen was not only aware of Savery's fire engine, but seems at some stage to have had a business connection either with Savery or his heirs. This might have been forced on him by the fact that Savery held a patent for an engine "impelled by the force of fire." No matter that Newcomen's engine was entirely different, the ignoramuses who dealt with patents in the British government would know only that Newcomen's engine was "impelled by the force of fire," too. Furthermore, some Newcomen experts claim that he was actually employed by Savery to build parts of his device; also

Dionÿsius Papin M.D.
Math. Prof. ord. ac
Reg. Soc. Lond. Soc.
Anno. 1689

that he built a Savery engine for himself in order to find ways of improving it.

Newcomen is also supposed to have corresponded with the erudite Dr. Robert Hooke, physicist, mathematician, and secretary of the Royal Society, and thus been steered onto the right track. This sounds a mite farfetched. First, Newcomen could barely write. Second, the exalted Dr. Hooke was one of the busiest scientists of all time and unlikely to enter into a discussion of physics, steam, and hydraulics with an unlettered West Country iron basher. It would be nice, however, to think that he had.

Newcomen had a partner, one John Calley, who did the rough work. Newcomen was the brains of the pair. The "fire engine" they contrived would, for the better part of the eighteenth century, do almost all of the mine-pumping in Britain. And more important, it would form the basis of James Watt's great invention.

The Newcomen engine of about 1712 was not very complicated. It consisted of a monstrous vertical cylinder inside which a piston slid. The piston rod stuck out of the top end of the cylinder which was wide open. The bottom end was enclosed, except for openings to admit steam and cold water. The piston rod was attached by a chain to a big wooden rocking beam to whose other end was attached—also by a chain—a water pump.

It worked like this: Steam from the boiler was let in below the piston as it rose upward toward the top of its stroke. (Having been pulled up by the downward stroke of the water pump; the steam did no pushing.) A spray of cold water was then injected into the steam, condensing it, and causing a vacuum which sucked the piston down with great force. This downstroke of the piston, transmitted through the wooden rocking beam, pulled the water-pump piston up. Upon discharging its gulp of water, gravity pulled the water-pump piston down, again raising the steam piston to the top of its stroke ready for the next cycle.

At first it took three men to work the various valves and the boiler. Later one man fired the boiler and a boy worked the steam-admission cock and the water-injection jet. A Newcomen engine didn't rev very fast—about fifteen strokes a minute was usual, during which time it might pump 160 gallons of water.

This meant that a considerable amount of valve twiddling was necessary each minute—too much, in fact, for a boy called Humphrey Potter. Humphrey figured out a system of strings and sticks whereby the engine would open and shut its own cocks. Which lazy ingenuity perhaps makes Master Potter the father of automation.

Simple as Newcomen's engine was, it still strained the technical competence of the eighteenth century. Boring a truly round hole was still an impossibility. Newcomen's pistons just didn't fit his cylinders. Luckily, the steam in a Newcomen engine had little pressure. Otherwise the rag, leather, and rope packing would have blown out at the first whiff of steam. The low-pressure, low-temperature steam also permitted ridiculously fragile boilers. The bottom of a boiler—the part over the fire—was made of copper. And *lead* was beaten into a beehive shape to form the upper parts of early boilers! Later, riveted iron sheets were used. Early cylinders were made of cast brass. Their bores, as they came from the molds, were too rough, so they were smoothed by hand rubbing with stones and sand. Imagine the knuckle-bruising labor! As casting methods improved, cylinders were made of iron, which was much cheaper than brass. A brass cylinder twenty-nine inches in diameter cost some £250 in 1725 (about $10,000 today). A few years later an iron cylinder cost but £20. Brass cylinders worked better, however, since they could be cast considerably thinner. A brass cylinder a third of an inch thick heated and cooled much more quickly than an inch-thick iron cylinder, thus requiring less coal.

By the mid-eighteenth century hundreds of Newcomen pumping engines were in use, some of them with cylinders more than six feet in *diameter.* Such huge engines consumed enormous amounts of fuel. This voraciousness didn't matter much in the coal mines of the north—in Yorkshire, Staffordshire, and Lancashire. But there was no coal in Cornwall and the cost of transporting the stuff by sea, and overland by wagon, was prodigious.

After Newcomen died in 1729, other men—notably James Smeaton—improved his atmospheric engine. Despite

Below: Thomas Savery's "fire engine" pumped water by means of a vacuum produced by condensed steam. It had neither piston nor cylinder. Right: H. Beignton's famous 1717 engraving of the Newcomen engine.

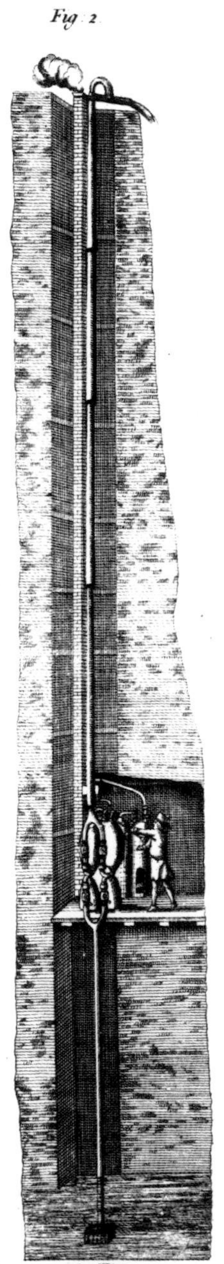

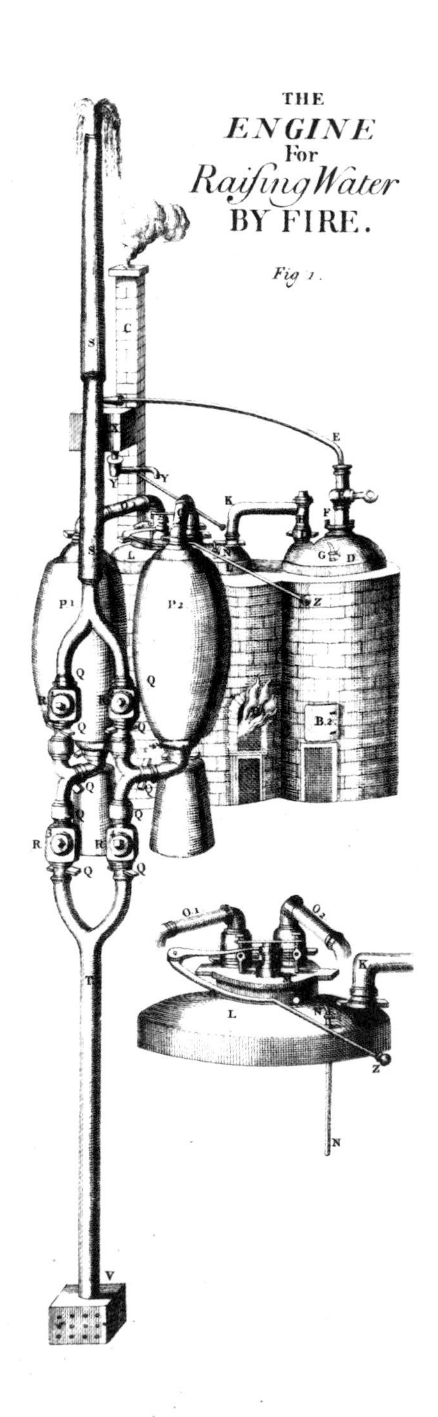

THE
ENGINE
For
Raising Water
BY FIRE.
Fig. 1.

Raising Water (with a power made) by Fire.

Right: James Watt pondering the poor operation of Glasgow University's model of the Newcomen engine. This nineteenth-century French illustration got the details of the engine model wrong. The actual model Watt worked on is shown opposite.

considerable increases in efficiency and economy it still was too extravagant of fuel, too slow and ponderous. Further improvement was needed and James Watt supplied it. He turned the atmospheric engine into a real steam engine.

James Watt was a Scotsman born in Greenock in 1736. At twenty-three he was already deeply interested in steam power. (The pretty myth about Watt first pondering on steam while watching a teakettle at the age of five is about as true as the story of George Washington and the cherry tree, which was supposed to have taken place at about the same time.) In 1759 Watt was already established in trade as an instrument maker, having spent a year in London learning to make quadrants, azimuth compasses, theodolites, and other mathematical instruments.

Upon his return to Scotland in 1756, Watt tried to set up shop as a mathematical instrument maker in Glasgow. The local guild concerned with such a trade, the Corporation of Hammermen, turned him down on the grounds that he was neither the son of a burgess nor a properly qualified apprentice of the borough.

Watt had, however, done a job for a professor at Glasgow University which impressed the academic types there with his competence. This consisted of cleaning and repairing a batch of instruments a man in the West Indies had left to the university in his will. The long sea voyage had rusted their steel, and their brasswork was covered with verdigris, but young Watt had very nicely refurbished them. The professors thereupon gave him a room at the university in which to carry on his instrument business, considering themselves lucky to have a young man who could repair scientific apparatus on the premises. Watt was lucky, too. For now he came into close contact with bright young scientific minds.

Watt's shop fascinated the professors. For he built not only scientific devices, but also fiddles, pipe organs, and guitars in order to augment his meager income. Among the academics who hung about was a youthful scientist, Jack Robison, who was a steam enthusiast. He even talked to Watt about such wildly futuristic ideas as wheeled carriages propelled by steam. Watt, a dour and conservative Scot even as a youth, became interested enough to try to reproduce some of Papin's experiments. Then he became busy with other matters and Robison went off with the British fleet dispatched to support Wolfe's attack on Montcalm at Quebec in one of the crucial battles of the French and Indian Wars. Steam locomotion had to wait.

When Robison returned to Glasgow in 1763 he found Watt no longer a rather timid and callow youth, but a self-educated man well grounded in engineering. Nor did he find him in his little room at the college. Watt had acquired a partner and was in his own premises in the town, where he made and sold "quadrants, Gunter's scales, compasses, microscopes, magic lanterns, etc., etc."

Now came a momentous turning point, not only for Watt, but for the world. Glasgow University owned a small model of a Newcomen atmospheric engine which refused to work. In desperation it had been shipped all the way to an instrument maker in London for repair. When it came back it still wouldn't work. Watt was asked to take a crack at it.

"I did so," said Watt later, "as a mere mechanician, but then began to study it seriously."

It was a lovely little model mounted in a nicely finished wooden frame. Its keg-shaped boiler was about nine inches in diameter. The piston had a two-inch bore and a six-inch stroke.

Said Watt: "By blowing on the fire it was made to take a few strokes." The boiler, obviously, was too small. It didn't make enough steam, and Watt calculated that about three-quarters of that steam was wasted, anyhow.

Watt realized that the cylinder couldn't be kept hot enough if it were going to be treated to a dash of cold water every time the engine made a stroke. Only a very hot cylinder—at least 212° F—would allow steam to enter at the beginning of the stroke without premature condensation. How to do this and still cool it off enough to induce the vacuum at the end of the upstroke?

Watt conducted many experiments with steam. He tried boiling water at varying pressures. He measured the volume of steam resulting from a quantity of water at atmospheric pressure and discovered that its volume increased about 1800 times. He found that water which had been con-

In 1765, JAMES WATT,
In working to repair this Model,
belonging to the Natural Philo-
sophy Class in the University of
Glasgow, made the discovery of a
separate Condenser, which has
identified his name with that of the
STEAM ENGINE.

Right: Glasgow University professors made Watt's shop a gathering place.
Below left: Watt's first crude engine incorporated a separate condenser.
Below right: Matthew Boulton.
Opposite: James Watt.

Right: Trevithick's high-pressure steam engine, 1805. Opposite: Watt's rotative beam engine as it now exists in the Science Museum, London. Note that a connecting rod, rather than a crank, drives the flywheel through sun-and-planet gears.

verted into steam could heat about six times its own weight of water (at room temperature) to the boiling point. This surprised him. He consulted Professor Joseph Black, one of the university men with whom he had become friendly. Black told him about his discovery of latent heat in 1761—that substances changing their physical state either give out or absorb heat at the moment of change. (For instance, ice absorbs heat as it melts into water, as does water when it boils into steam.)

Watt said later, "I thus stumbled upon one of the material facts by which this beautiful theory [latent heat] is supported."

Watt chewed on the problem, worrying about it as he wandered about Glasgow one Sunday. Suddenly the solution hit him. As he told an engineer friend years later: "It was in the Green of Glasgow. I had gone to take a walk on a fine Sabbath afternoon. I had entered the Green by the gate at the foot of Charlotte Street—had passed the old washing house. I was thinking upon the engine at the time and had gone as far as the Herd's house when *the idea came into my mind, that as steam was an elastic body it would rush into a vacuum, and if a communication was made between a cylinder and an exhausted vessel, it would rush into it, and might be there condensed without cooling the cylinder.* I then saw that I must get quit of the condensed steam and injection, if I used a jet as in Newcomen's engine. Two ways of doing this occurred to me. First the water might be run off by a descending pipe, if an offlet could be got at the depth of 35 or 36 feet, and any air might be extracted by a small pump; the second was to make the pump large enough to extract both water and air. . . . *I had not walked further than the Golf-house when the whole thing was arranged in my mind.*"

Watt must have been itching to get to his shop to try out his idea. But it was Sunday and, although Watt wasn't much hampered by religious foibles, it just wasn't seemly, two hundred-odd years ago, to work on the Sabbath.

The next morning he set to. Hurriedly he slapped together a crude model. (It still exists in London's Science Museum.) A brass syringe serves for a cylinder and soldered-up tin-plate as a condenser; it even incorporates his wife's sewing thimble.

Watt's little model not only had a separate condenser from which the water was removed by a small engine-driven pump, it also had a cylinder covered at both ends (unlike Newcomen's open-at-the-top cylinder) and a piston rod which protruded through a steam-tight stuffing box at the bottom end.

Watt attached an eighteen-pound weight to the bottom of the piston rod. The crude little engine proved, by lifting the weight, that it could perform work.

The invention was whole inside James Watt's head now, but it would take him a long time, and much worry and labor and money, before he could work out the many details of a practical model he could patent.

Some of these details were obvious to Watt immediately: the separate condenser, the air and water pumps, the use of animal fats and vegetable oils (mineral oil was as yet unknown) to seal the piston in the cylinder. (Newcomen's engine had relied on water lying on top of the piston to help seal it.) Most important was Watt's realization that he had to keep the cylinder hot. He envisioned insulating it with wooden lagging or, better still, surrounding it with a steam-filled jacket.

Now began poor Jamie Watt's travail. He rented an unused pottery and tried to hire workmen who had the mechanical ability to build his engine. Watt himself was a superb mechanic, but he was used to working small with delicate tools. He had no experience "in the practice of mechanics *in great*," as he put it. Mechanics who could build large, heavy machinery with any degree of accuracy were nonexistent. The only workmen to be had were blacksmiths and tinners experienced only in bashing out crude and clumsy ironwork for sailing ships and wagons, or pots for cooking or brewing. And many of these were drunken, blundering incompetents at their own rude trades.

Watt's first cylinder was about half a foot in diameter and had a two-foot stroke. No one in Glasgow could bore such a cylinder. It had to be roughly hammered into shape.

This cylinder seems to have been thrown on the scrap heap. The cylinder of Watt's first engine was cast in block tin in the foundry of the Carron Iron Works, the same

founders who cast the "carronades," those famous cannon used to arm Britain's fighting ships of the time. The Carron Iron Works had recently been established by a Doctor John Roebuck, who also had other interests—gold and silver refining, chemicals, mining.

Watt had met Roebuck through his friend Dr. Black, who had now and then been slipping a few pounds to Watt to help him along with his experiments. Watt, who made very little out of his instrument business, needed much more cash. Roebuck undertook to finance the engine in exchange for two-thirds of the profits.

This paid for the engine work but did not help Watt make a living. Among other jobs he took on work as a surveyor of canals. Furthermore, Watt, like so many other inventors, didn't stick to his original design. He continually tried improvements. He even tried working out a rotary device—a sort of Wankel steam engine.

Watt was no longer in his pottery cellar. He was erecting his engine in a workshop on the grounds of Roebuck's estate at Kinneil. It was July, 1769, before it was at last put together—six months after a patent for it had been granted. The steam boiler was set over its brick furnace, the huge wooden rocking beam was hoisted into position. The tin-plate condenser and its pumps were set into a box-like frame. It was the wood-insulated cast-iron cylinder that worried Watt. It was, by today's standards, a fairly big one: eighteen inches in diameter with a five-foot stroke. And "it was," Watt said, "a clumsy job." Its piston was no better. Tradition has it that Watt tried greased cardboard, linseed oil, rags, tow, old hats, and horse dung to fill up the spaces between the cylinder wall and the piston. The engine worked, but lamely and far below Watt's expectations. It spat steam from a dozen joints. Its piston leaked.

Watt was despondent. He had never been a robust, carefree type. Poor health had long plagued him and he had suffered from terrible migraine headaches all his life. Now he was in the depths of depression. Nor was Roebuck too happy while Watt fussed with the engine, altering this, improving that, then going off surveying. Roebuck tried to get him moving. "You are letting the most active part of your life insensibly glide away," he wrote. "A day, a moment ought not to be lost." In letters to Roebuck Watt sadly agreed. "Much contrived and little executed," he said. And, "of all things in life there is nothing more foolish than inventing."

Things got worse. Roebuck had overextended his speculative investments. A business depression caught him. In 1773 he went broke. Watt had spent much more than the money—£1,000—advanced to him by Roebuck, and it was agreed that the ailing engine would become Watt's property. Roebuck still retained a two-thirds interest in the patent which now seemed valueless. At this low point came another disaster. Watt's wife died, and he was left with two small children, both under six, to care for.

Watt could not have realized it then, but his ill luck had bottomed out. His fortunes were about to rise and, with minor dips, keep rising until he would more than realize the great expectations he had once had for himself and his "fire engine." For a unique and brilliant man became intrigued with Watt's engine.

This was the great Matthew Boulton, a wealthy industrialist at a time when industrialists were a rare breed. His factory at Soho, near Birmingham, where he carried on a large business in high-class silver stampings, sword hilts, watch chains, buckles, buttons, ormolu (some of which is now prized by collectors), was a remarkable undertaking for the eighteenth century, when most such production was piecework by journeymen serving under masters in tiny shops. The building in which more than six hundred employees worked looked more like a Georgian palace than a factory.

Boulton and Roebuck were friends. In 1768, when Watt took a trip to London to talk to politicians about the patent, Roebuck suggested that he stop off at Soho to meet Boulton and his associate, Dr. Benjamin Small. Boulton and Watt hit it off immediately. Further, they were excited by Watt's "fire engine" as a power source for Boulton's factory, which was run by water power.

Boulton and Small suggested joining Watt and Roebuck in the steam-engine venture, but Roebuck resisted the idea. It wasn't until 1773, when Roebuck at last went

Opposite:
After his retirement, Watt spent years in this attic trying to devise a machine for copying sculpture.

bankrupt, that Watt could join Boulton. Roebuck owed Boulton a considerable sum and to liquidate this he ceded all his rights in the patent.

In May, 1774, Watt and his children, household goods, and tools left Scotland for Birmingham. The Kinneil engine had been packed up and shipped to Soho months before. Now he was welcomed by his new partner, Boulton, who was rich, shrewd, influential, and as firm a friend as any poor, thirty-eight-year-old Glaswegian steam-engine inventor could possibly desire.

Watt applied himself to his engine. Soon he was able to write to his father: "The fire engine I have invented is now going and answers much better than any other that has yet been made." This improvement was due in large part to a new cylinder made by John Wilkinson of Bersham. Wilkinson, a well-known ironmaster and a friend of Boulton, had recently received a patent for a boring mill which produced cylinders whose bores were almost truly round for their whole length.

The engine worked but Boulton was worried. The patent now had but eight years to run. Watt, therefore, journeyed to London and asked Parliament for a twenty-five-year extension. Boulton helped with the lobbying and the extension was granted. This virtual monopoly until the turn of the nineteenth century was later much inveighed against—and perhaps rightly—as stultifying to progress. In June a new firm was born, known by one of the great company names of all time: Boulton & Watt.

Boulton, mover and shaker, immediately brought in orders—one for a large pumping engine with a 50-inch-diameter cylinder for the Bloomfield Colliery in Staffordshire, the other for John Wilkinson's ironworks. Although Watt was nervous about plunging into the construction of engines for sale before more experimenting with the Kinneil engine, Boulton bade him charge ahead but warned, "Don't let the engine make a single stroke until certain that it will work without a hitch, and then, in the name of God, fall to and do your best."

That big brute of a Bloomfield engine, with a piston more than four feet in diameter, was a thumping success. As was the Wilkinson engine. (Wilkinson had indeed forged most of the big parts for both engines.) Both used only about twenty-five per cent of the coal a Newcomen engine burned for the same amount of work. Orders poured in, mostly from those everlasting water-fighters, the Cornish mine operators.

These engines were, as yet, single-acting. The steam entered the cylinder on only one side of the piston—on its upper side while it was on its downward stroke. Its upward stroke was a pull-up caused by the downstroke of the pump at the other end of the rocking beam.

In 1781 Watt patented his double-acting engine, which, by means of the now familiar steam chest and slide valves, shoots steam in first at one side of the piston, then at the other. At about the same time Watt introduced the important idea of the "cutoff."

Watt at first used his steam at low pressure—only slightly above atmospheric pressure. He then found that if he caused it to enter his cylinder at considerably higher pressure he could stop admitting it when the piston had traveled only part way and allow the natural expansion of the steam to push it the rest of the way. By thus "cutting off" the steam a great saving in fuel was possible. Watt discovered that he could cut off the steam when the piston had done only a quarter of its stroke. Watt thereupon designed a mechanism which not only slid the valves open and shut, but also operated the cutoff. The increased efficiency was enormous.

It was now late in the eighteenth century. The Industrial Revolution was in full acceleration. Cotton mills, iron-rolling mills, blast furnaces, forges, and a dozen burgeoning new industries needed a new kind of power to supplant the old water wheel. A steam engine with a rotative capacity was needed, not merely one which could pump mines dry.

Watt seems to have been wrongheaded in his approach to this problem. Instead of converting the in-and-out motion of the piston rod to a round-and-round motion by means of a crank, a method long used in treadle-operated devices such as lathes and potter's wheels, he devised a peculiar arrangement of gears. A fixed gear at the end of the rod (which formerly had been the pump rod in pumping engines) engaged another gear at the center of a big flywheel.

Right: Jonathan Hulls.
Opposite: Hulls's Newcomen-engined stern-wheeled towboat of 1736.

One gear revolved around the other, giving the name "sun and planet" to the arrangement.

This not-too-logical method was, however, caused by a patent problem. Watt had indeed been constructing an engine with a crank incorporating a means for getting over dead center. One of his workmen, one Dick Cartwright, talked about it in a pub. An eavesdropper heard him and Watt soon found that the idea for using a crank in combination with an engine had been patented. Later Watt engines, however, did use cranks and flywheels.

Watt also invented the centrifugal governor, that little device with whirling nickel-plated balls which, forty years ago, so fascinated small boys watching steam engines at work (something missed by modern small boys who have nothing but diesel engines to admire). When a steam engine ran too fast, centrifugal force pulled the spinning metal balls outward and upward on the rod to which they were attached by levers. The upward movement shut the throttle. If engine speed faltered, the balls dropped, opening the throttle.

If the Industrial Revolution had been moving ahead without the steam engine to power it, it now went roaring headlong into the new capitalist-industrial nineteenth century. And after some painful financial problems, the fortunes of Boulton and Watt soared with it.

Oddly, although Watt had dreamed as a young man of powered vehicles and ships, he later took a dim view of such uses for his engine. True, he had included such an application in his patent of 1782. But this was to placate a valued assistant, William Murdock, who was enthusiastic about steam-propelled road vehicles. In fact, Murdock, to Watt's annoyance, had a model nonpassenger-carrying steam carriage on the road as early as 1784 (when it terrified a village parson in Cornwall).

James Watt, rich and full of honors, retired in 1800. Living the quiet existence of a country gentleman, he did not give up inventing. In the garret of his big country house he labored for years on a machine for copying sculpture. He died in 1819, eighty-four years old.

Other men had already applied the steam engine to ships and railways.

Steam on the Water

Robert Fulton did not invent the steamboat. The idea of using a steam engine to drive a vessel goes back to at least 1707, when Denis Papin tried turning paddle wheels with his primitive engine. Using ratchets to turn the paddles, he seems to have actually succeeded in operating his small vessel. However, when he brought his steamer down the Fulda River in Germany to the Weser for a demonstration, its paddle wheels were not connected to the engine but were rotated by his crew. At Munden, German watermen, fearful of competition, boarded the boat and destroyed it.

In 1729 a Dr. John Allen, in England, patented a steamboat which was to be propelled by a jet of water forced sternward by engine-driven pumps. Twin Newcomen-type atmospheric engines were to do the pumping. Again, when this vessel was tried out on a canal, manpower not steampower worked the pumps. In 1736 Jonathan Hulls managed to get a patent for a Newcomen-engined sternwheeler. Old prints show it towing a big sailing ship. This granddaddy of all tugboats was described in Hulls's patent as an "invention of a machine for carrying ships and vessels out of or into any harbour or river against wind or tide."

The Comte Claude de Jouffroy d'Abbans, in France, was first with a genuinely practical steamboat, which just missed being commercially successful. He launched his first vessel in June, 1778. It was small, only forty-three feet long, with peculiar paddles that flapped like a duck's feet. A twin-cylindered Newcomen atmospheric engine drove the paddles through chains. Too little power was developed and the boat failed to perform very well when it was tried on the Doubs River. But le Comte was undismayed. Five years later he tried again with a new vessel, the famous *Pyroscaphe,* which was built near Lyons and fitted with machinery built by Frère Jean et Cie. A double-acting steam engine, whose single cylinder was inside the boiler, drove two big 13-foot paddle wheels via ratchets. The paddles, or floats, were so pivoted that they

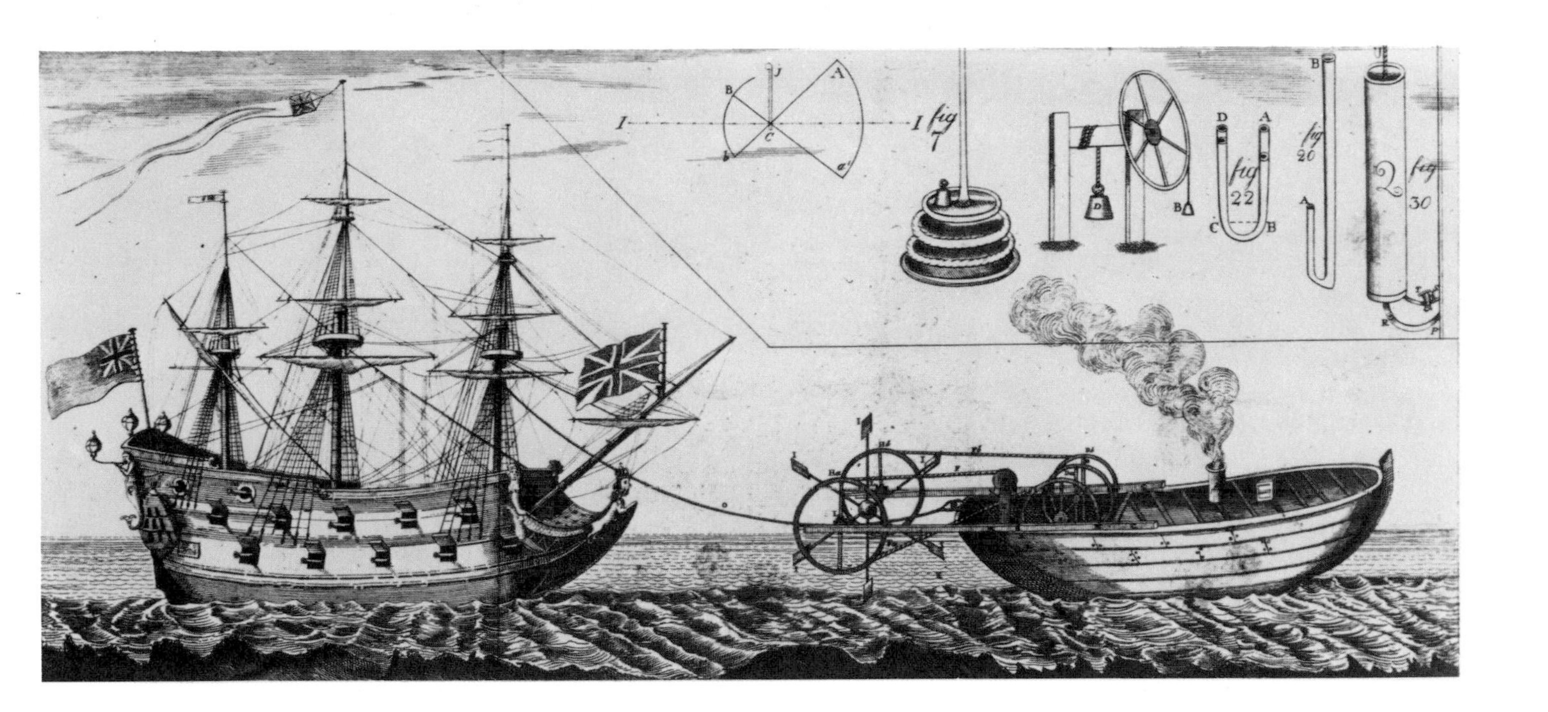

caused no resistance when they first entered the water and wasted no power lifting water on their upstroke. The cylinder was a big one, with a 25-inch bore and a 77-inch stroke, but Jouffroy's boiler was incapable of making enough steam for a long, continuous run. The *Pyroscaphe* was, however, very successfully demonstrated in July, when she was said to have breasted the current of the Saône River at about four knots. Despite this, the conservative government of Louis XVI was blind to the important breakthrough Jouffroy had achieved and gave him no financial help. It wasn't until after Waterloo, after the traumas of the Revolution and Napoleon, that Jouffroy was able to launch his fine big steamboat, the *Charles-Philippe,* at Bercy in 1816.

Inventors have traditionally suffered poverty, indifference, hypocrisy, and fraud. But the saddest story of an inventor's evil luck is that of John Fitch. This Connecticut Yankee built several successful steamboats, without, however, gaining recognition, and eventually, penniless and in despair, he killed himself.

Born in 1743 on a stony farm at Windsor, Connecticut, he was apprenticed by his bigoted, tight-fisted father to a clockmaker who not only cheated him, but taught him noththing beyond the rudiments of brass-working. Later he married a shrewish, sluttish wife whom he left to become a wanderer. He eked out a living by doing odd jobs—silversmithing, surveying, even serving as a sutler for Washington's army at Valley Forge.

Fitch was one of those rare, natural craftsmen who could turn his ingenuity to almost any mechanical task. But in 1782, to escape from the relentless poverty in which such work kept him, he headed west of the Alleghenies to stake out land for himself. With the usual Fitch luck, he was captured by Indians in the pay of the British and almost lost his scalp.

It was after the revolution, in 1785, that he first became imbued with the idea of building a steamboat. All he knew about steam was learned from an engraving of the model of Newcomen's engine at Glasgow University (the one Watt had repaired) which he found, *after* he got the idea for steam propulsion, in Martin's *Philosophia Britannica.* He was, he said later in his autobiography, "considerably chagrined." Never having heard of it before, he thought that the idea of the steam engine was his alone, although three Newcomen-type engines were in fact operating in the United States.

He immediately built a two-foot brass model of his steamboat. First, he devised a propulsion system using a long double chain with wooden paddles attached to it, like fins. Alternatively he designed circular paddle wheels. He tried his boat in a stream, "working the paddles by hand" (we can't imagine how). He was satisfied that his ideas worked and set out to get backing to carry his invention further. He sold maps of the western wilderness that he had drawn and engraved (and printed on a cider press). Then he gathered testimonials from various bigwigs to bolster a petition for government aid, but a congressional committee ignored his request. He then showed his model to the Spanish minister. "Would Fitch, if aided, give exclusive rights to His Most Catholic Majesty?" Fitch, the patriot, spurned the offer and lived to regret it.

He approached Benjamin Franklin, who amiably allowed that Fitch's steamboat would be a fine thing while doing nothing to advance it. George Washington was noncommittal, the Virginia Assembly, Tom Paine, and Thomas Jefferson wished him well—no more. (Paine and Jefferson did, however, buy maps.)

But huzzah! The New Jersey Legislature voted him "the sole and exclusive right of constructing, making, using, and employing, or navigating, all and every species or kind of boats . . . impelled by the force of fire or steam, on all the creeks, rivers etc. within the territory or jurisdiction of this state" from March 18, 1786, for fourteen years.

Fitch's troubles were just beginning. Monopoly or no, he still needed money to build a full-size steamboat. To this end he formed a stock company which raised $300 for building a steam engine and fitting it into a boat.

Luckily, Fitch fell in with a fellow mechanical genius, Henry Voight. Although neither had ever seen a steam engine, they managed between them to devise a double-acting engine, a type James Watt and William Murdock had developed only after many years of work. Amazingly, both engines were completed the same year. Fitch's engine, a small

Right: Model of Frenchman Jouffroy d'Abbans' 1794 steamboat.
Opposite, top: John Fitch's steamboat at Philadelphia, about 1787.
Bottom left: Fitch's drawing of his 1790 engine.
Bottom right: A romanticized French version of Fitch's suicide.

one with a three-inch-bore cylinder, was installed in a skiff fitted with his chain-and-paddle device. Launched on the Delaware, it failed. The chain-and-paddle system wouldn't move the boat. Fitch then installed a weird means of propulsion that looked like a crew of Indians wielding paddles in a canoe—but without the Indians. Happily, this worked.

Fitch told his stockholders that a cylinder of 12-inch bore would propel a vessel of "20 tons burthen 10 Miles per Hour, if not 12 or 14. . . ."

The stockholders' pockets remained shut. Fitch, in a frenzy, hied himself from legislature to legislature and succeeded in getting monopolies to manufacture and operate steamboats from Delaware, New York, and Pennsylvania.

The stockholders relented and construction was started on a boat forty-five feet long. The hull was to cost $50. If James Watt had had trouble getting decent workmanship on his engines, imagine the problems in mechanically backward America. And imagine the jeers when waterfront idlers watched a seven-thousand-pound brick furnace being built on the boat's deck. In spite of terrible difficulties and the screams of the stockholders when the machinery cost £530, the boat was ready in June, 1787.

In its first trials, with the engine clanking and leaking steam, Fitch's boat moved against the current of the Delaware at two-and-a-half miles an hour.

The famous painter Rembrandt Peale described the trials: ". . . on the deck was a small furnace and machinery connected with a coupling crank projecting over the stern to give motion to three or four paddles resembling snow shovels which hung into the water. When all was ready the power of steam was made to act . . . the paddles began to work, pressing against the water backward as they rose, and the boat to my great delight moved against the tide without wind or hand."

By the time Fitch finally had a boat good enough to put into commercial service between Philadelphia and Trenton he was dead broke, starving and ragged, and living without payment in a public house owned by a friendly widow. He had sold his stock to raise money for experiments and had gone through a bad time fighting the patent claims of one James Rumsey, whose water-jet-propelled steamboat had been backed in part by George Washington and Benjamin Franklin.

In 1790 advertisements appeared in the *Federal Gazette* and the *Pennsylvania Packet:* "The steamboat is now ready to take passengers, and is intended to set off from Arch Street ferry, in Philadelphia, every Monday, Wednesday and Friday for Burlington, Bristol, Bordentown and Trenton."

From May to September, Fitch's steamboat traveled almost three thousand miles. It ran eighty miles in a day and at 8 mph against the current. It was much faster than Fulton's *Clermont* of seventeen years later.

But the line ran at a loss and the stockholders refused to put up more cash. Work started on another boat, the *Perseverance,* was abandoned by the company, although Fitch still labored hopelessly on it.

A year later Fitch's hopes rose once more. Aaron Vail, who had seen Fitch's successful runs on the Delaware and was U.S. Consul at Lorient, France, petitioned the French government for the exclusive privilege of building steamboats in France. The rights were to be shared in a complex way by Fitch, the Steamboat Company, Voight, and Vail. With a letter of credit from Vail, Fitch traveled to France to build a new boat. But France was in the throes of revolution and nothing happened. Fitch then crossed to England with no better luck. He returned to Boston in steerage as an indentured laborer, working out his passage after landing in the United States.

Some years later, in 1796, he ran a tiny steamboat with a wooden cylinder on Collect Pond (now the site of the Tombs prison), in New York City. One of his passengers was Robert Livingston, who ten years later backed Robert Fulton. The reports of this are, however, vague and dimmed by time. In any case, it is hard to understand why Fitch went back to fussing with such a tiny and primitive craft.

He traveled west again that year to claim thirteen hundred acres he had staked out in 1782. But now the wild lands had become squatters' farms and he did not succeed in reestablishing his ownership. For a time he experimented with a model of a locomotive—a quite practical one, too. Then he turned to drink, paying for each pint with useless land

JOHN FITCH
1786

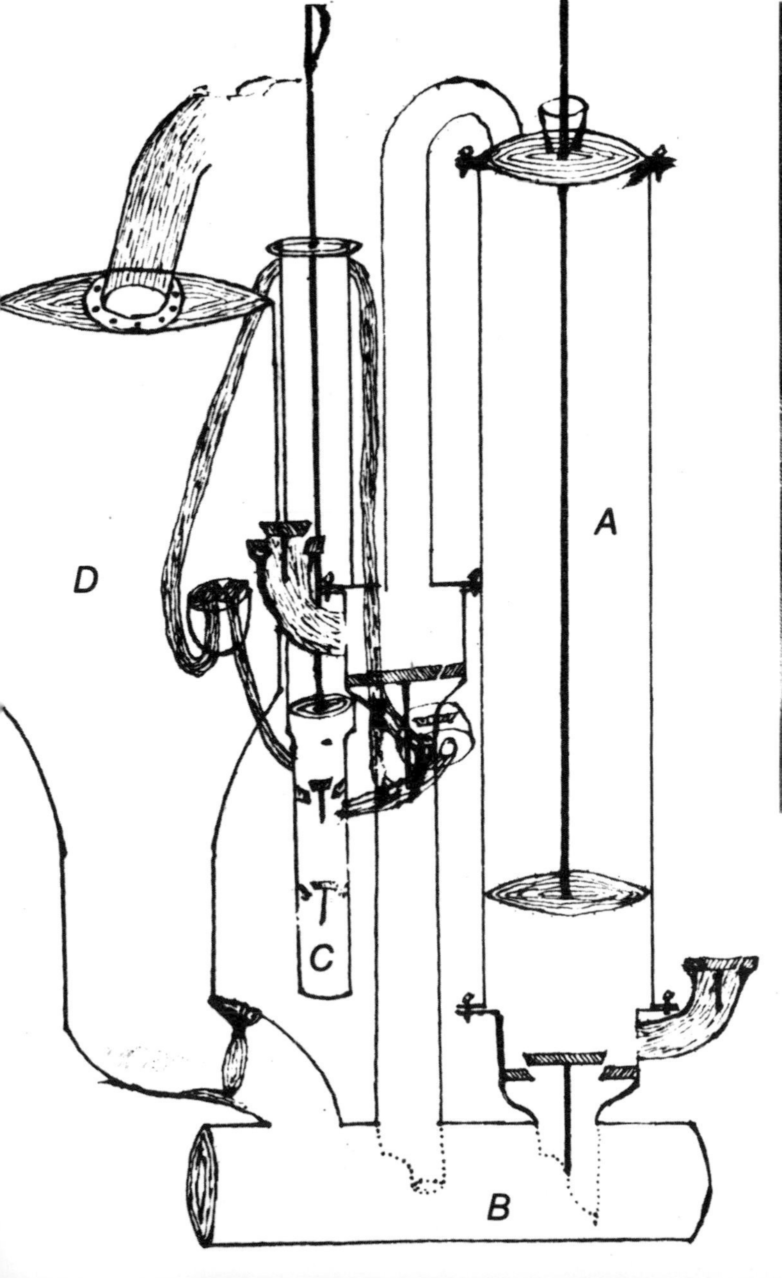
A
B
C
D

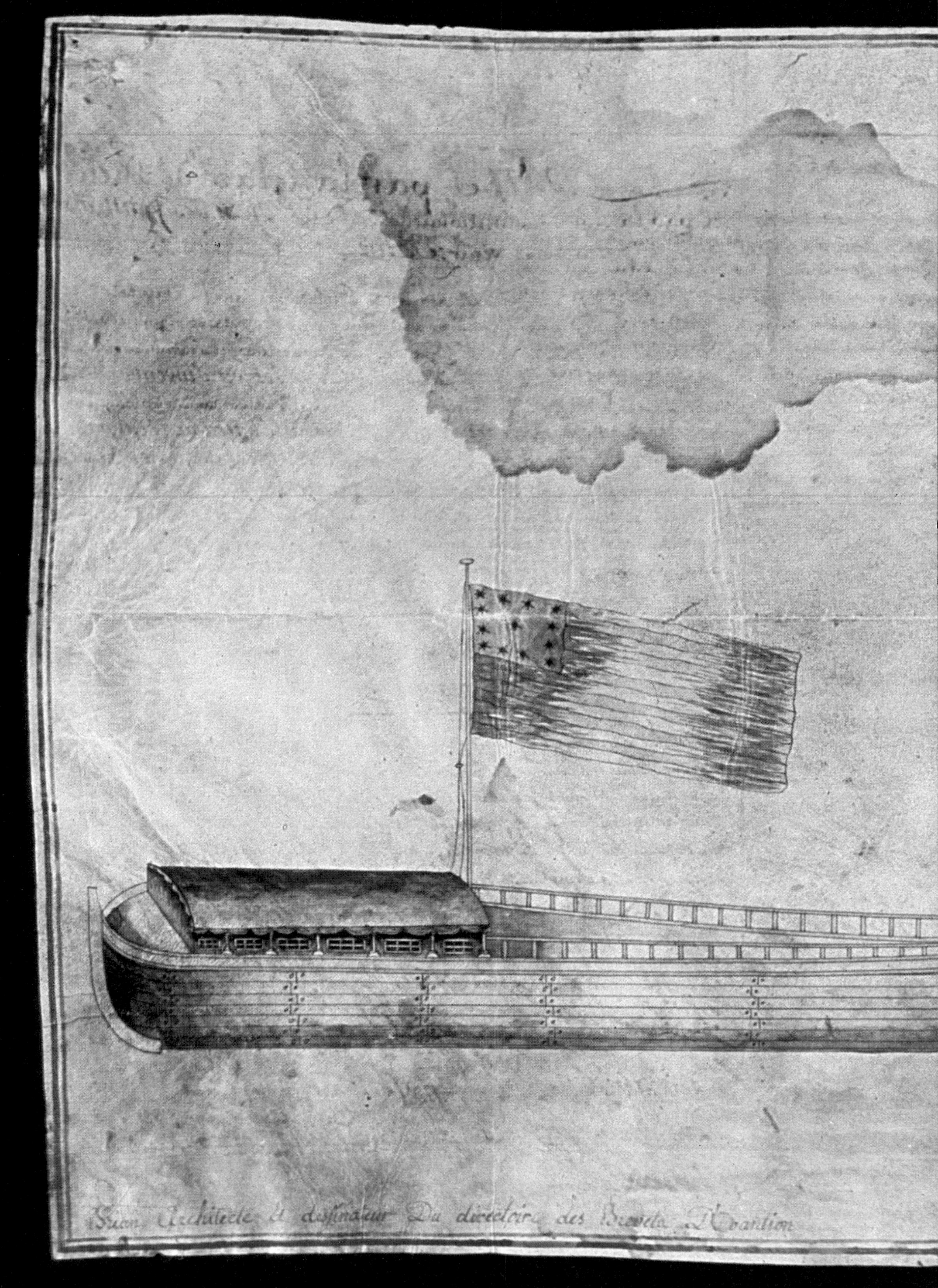
Simon Architecte et dessinateur Du directoire des Brevets d'Invention

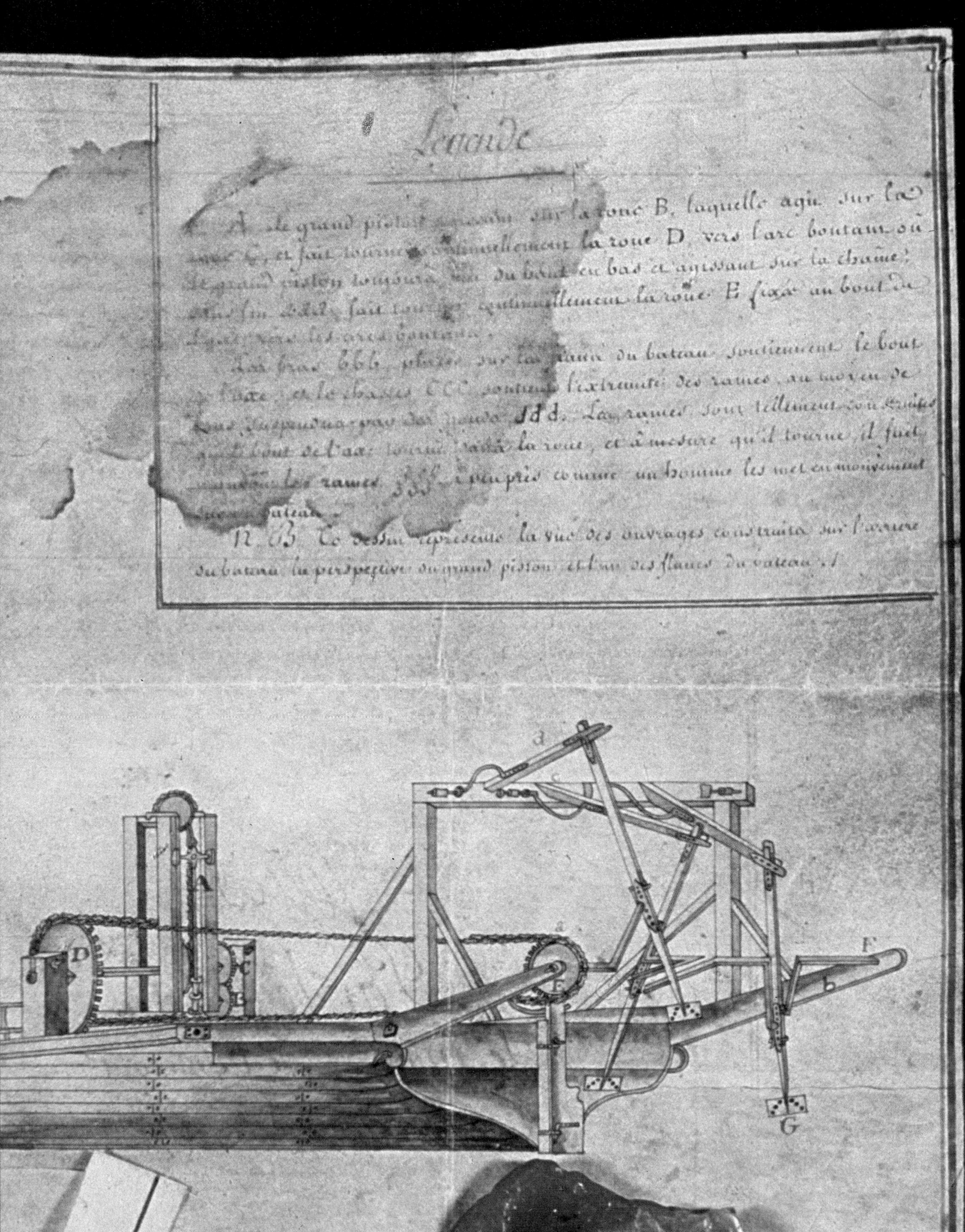

Legende

A le grand piston [illegible] sur la roue B, laquelle agit sur la [illegible] C, et fait tourner continuellement la roue D vers l'arc boutant où le grand piston toujours [illegible] du bout en bas et agissant sur la chaîne [illegible] fait tourner continuellement la roue E fixée au bout de [illegible] les arcs boutants.

Les bras bbb placés sur les flancs du bateau soutiennent le bout de l'axe et les chasses ccc soutiennent l'extrémité des rames, au moyen de [illegible] suspendues par des [illegible] ddd. Les rames sont tellement construites que le bout de l'axe tourné [illegible] la roue, et à mesure qu'il tourne, il fait [illegible] les rames fff à peu près comme un homme les met en mouvement [illegible] bateau.

N. B. Ce dessin représente la vue des ouvrages construits sur l'arrière du bateau, la perspective du grand piston et l'un des flancs du bateau.

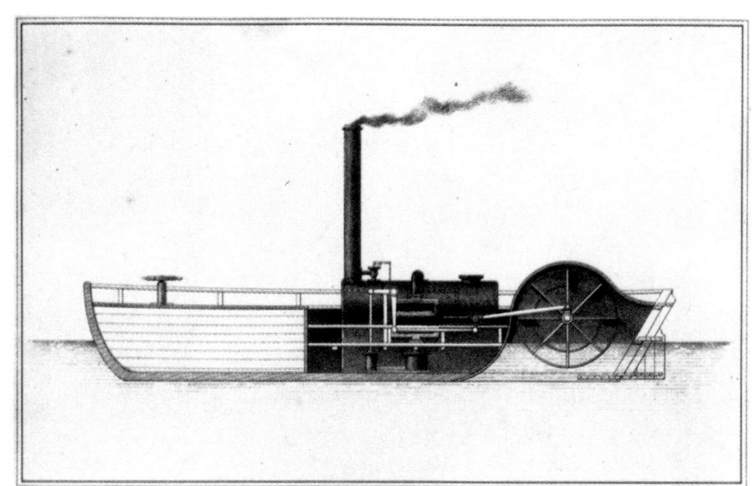

Preceding pages: 1791 drawing of the Fitch steamboat, for submission to the French patent office. Opposite, top: Patrick Miller's 1788 twin-hulled, Symington-engined steamboat. Middle left: Symington engine. Middle right: The *Charlotte Dundas.* **Bottom: Model of John Stevens' steamboat.**

warrants. He died a suicide from opium pills and whiskey in 1798.

By now, others had been working on steamboats, but none had surpassed Fitch.

In Britain an Edinburgh banker named Patrick Miller had been experimenting with twin-hulled boats with manually turned paddle wheels for use on canals. His son's tutor suggested that steam power might be better than man power and he turned to William Symington, who had patented a steam engine (in fact, an infringement of Watt's) in 1787. The first trials were fairly successful, the twin-hulled boat reaching some five miles per hour on Dalswinton Loch on Miller's estate in Dumfriesshire, Scotland, despite Miller's use of an archaic system of ratchets to attain rotary motion. A later twin-hulled Miller boat with an engine built by the Carron Iron Works reached a speed of seven miles an hour on the Forth and Clyde canal. After unsuccessfully trying to interest the admiralty, Miller, whose interest doesn't seem to have extended beyond canal navigation anyhow, gave up on steamboats.

Symington was again involved with a steamboat when Thomas, Lord Dundas of Kerse, one of the governors of the Forth and Clyde canal, wanted to try steam for towboats. Symington was commissioned to build a towboat and the highly successful *Charlotte Dundas* resulted.

The *Charlotte Dundas* had its paddle wheel hidden under a peculiar recess near its stern. This resulted in an odd double stern which required twin rudders. The boiler and the engine lived side by side. The engine's single cylinder had a bore of 22 inches, and a four-foot stroke. The piston rod was connected to a crank on the paddle-wheel shaft.

In March, 1802, the *Dundas* showed what it could do. Against a headwind so strong that no other vessels (towed by horses) could move that day on the Forth and Clyde canal, she towed two seventy-ton canal boats 19½ miles in six hours. But the canal owners, fearful of damage to the canal's banks by the wash from steamers, decided to stick with horses.

The *Charlotte Dundas* was left lying in a creek until she was broken up in 1861. Besides proving what a steamboat could do, she was important in another way. For Robert Fulton not only examined her with great care in 1804, he also managed to get Symington to take him aboard for a ride.

Still another American steamboat man active in the eighteenth century, somewhat ahead of Mr. Fulton, was John Stevens of Hoboken, New Jersey. Unlike poor, illiterate, ragged John Fitch, Stevens was a wealthy, educated dandy with close ties to the upper-crust Tory establishment. Yet when the revolution came, he deserted his pro-British class and took a commission in Washington's ragtag army. In 1782, shortly after the war ended, he married a famous and wealthy beauty, Rachel Cox. Rachel, who always called her husband "Mr. Stevens," was not devoid of humor. In bed one morning, Stevens woke with an idea for a new engine and sketched it with a finger on her back. She wakened and tried to shake off this annoyance. "Hold still!" growled Stevens, "don't you know what figure I am making?" "Yes, Mr. Stevens, the figure of a fool." But Colonel Stevens was no fool. He was thinking out a most practical steamboat.

Stevens owned a thousand-acre estate across the Hudson from New York in a place with the Indian name of Hopaghan Hackingh, now Hoboken. Fed up with the unreliability of the rowdy ferrymen who daily rowed him to and from the city, he bought the ferry monopoly, intending to make it a profitable business. Steam might be one way to do it.

Stevens' brother-in-law, that rich mover-and-shaker, the politician Robert Livingston, was interested in steam propulsion, and had paid close attention to Fitch's work. Both he and Stevens had seen Fitch's steamers in action on the Delaware, and were sharp enough to realize that they were seeing great ideas poorly executed by men much too short of money to do a proper mechanical job.

Stevens, inspired by Fitch, started his own experiments with steam. Livingston bought up the monopoly patent which the State of New York had granted to Fitch: the sole right to navigate the Hudson by steamboat, provided that a regular service was operated between New York and Albany —a condition Fitch could never fulfill.

A partnership resulted. Stevens was the inventor, Livingston the businessman, and one Nicholas Roosevelt the mechanic.

Right: John Stevens.
Opposite, top: Stevens' ***Phoenix,*** **the first ocean-going steamer.**
Bottom: Stevens' 1804 engine and twin screw propellers.

Stevens experimented. He decided that high-pressure steam from more efficient boilers was the way to increase power. Things went slowly but Stevens did succeed in running small, scow-like steamboats on the Hudson. These little twenty-footers puffing between New York and Hoboken were a common sight in the late 1790s. But Livingston was itching to get going faster. The time for which they would enjoy their monopoly was running out and Stevens still had no commercially feasible steamboat for the Albany–New York run.

As the eighteenth century ended, Livingston was rushed to France by President Jefferson to buy the Louisiana Territory from Napoleon. There he met Robert Fulton, who at that time was involved in trying to sell Napoleon an idea for a submarine to break the British blockade. Livingston, impressed with Fulton, saw in him the very man who could build a steamboat practical enough to fulfill the charter. But this meant terminating the relationship with Stevens. Livingston did his best to let Stevens down easy. He mentioned Fulton in a letter to Stevens but refrained from saying anything about Fulton becoming his new partner—which would, of course, make Stevens a new competitor.

By 1804, when Livingston returned from his dealings with Napoleon, Stevens had not only perfected a new high-pressure water-tube boiler but also a small vessel, the *Little Juliana,* which, instead of the paddle wheels he had used earlier, was driven by twin screw propellers. A New Yorker of the day said: "As we entered the gate from Broadway, we saw . . . a crowd running towards the river. On inquiring the cause, we were informed that 'Jack' Stevens was going over to Hoboken in a queer sort of boat. On reaching the bulkhead . . . we saw lying against it a vessel about the size of a Whitehall rowboat, in which there was a small engine *but no visible means of propulsion.* The vessel was speedily under way."

When Livingston told Stevens that Fulton was going to build a steamer for him which would carry out the terms of the monopoly, he also allowed as how Fulton was further along in perfecting such a vessel. Stevens would admit no such thing. Nor would he accept a partnership in the new venture. In Stevens' view, Livingston and he already had a binding agreement for Stevens and Roosevelt to build the boat.

For several years, Livingston and Stevens fussed and fumed in a long but friendly-seeming correspondence over legalities. Each continued his own way. Stevens had a steam ferry plying the Hudson a year before Fulton's *North River Steamboat* (later called the *Clermont*) made its famous run. But when that happened Stevens decided to stop bucking Livingston and the New York money men and politicians in cahoots with him. The steamboat *Phoenix,* which Stevens had built for the Hudson River trade, he sent to Philadelphia to establish a line on the Delaware. The *Phoenix* had to take to the open sea and round Cape May to reach the Chesapeake and thus became the first ocean-going steamer. A sailing vessel was sent along with it to give aid if needed. But a gale blew the schooner out to sea, while the little *Phoenix* churned through the big waves to Philadelphia. The schooner came in three weeks later.

Robert Fulton was born the year that James Watt took out his patent for the separate condenser steam engine—1765. Born on a farm in backwoods Pennsylvania, he had a certain native sharpness—an eye for the main chance. Before he was twenty he was a painter of miniatures in Philadelphia and had already managed to become acquainted with people who mattered, among them Benjamin Franklin. Although he had no great talent as a painter he had great ambitions. He considered the opportunities of Philadelphia too narrow, and spurred by the success of another American painter, Benjamin West, he set his sights on London, where West was one of the stars of the Royal Academy. Franklin gave Fulton a letter to West and the young artist set forth for England.

West took Fulton on as a pupil and in a few years succeeded in turning him into a quite mediocre painter of portraits, at which pursuit he almost starved to death.

But Fulton had talent. He had certain minor abilities in engineering, he was an ingratiating talker, and he had that enviable knack of being readily accepted by important and influential men.

STEAM BOAT PHOENIX
MOSES ROGERS

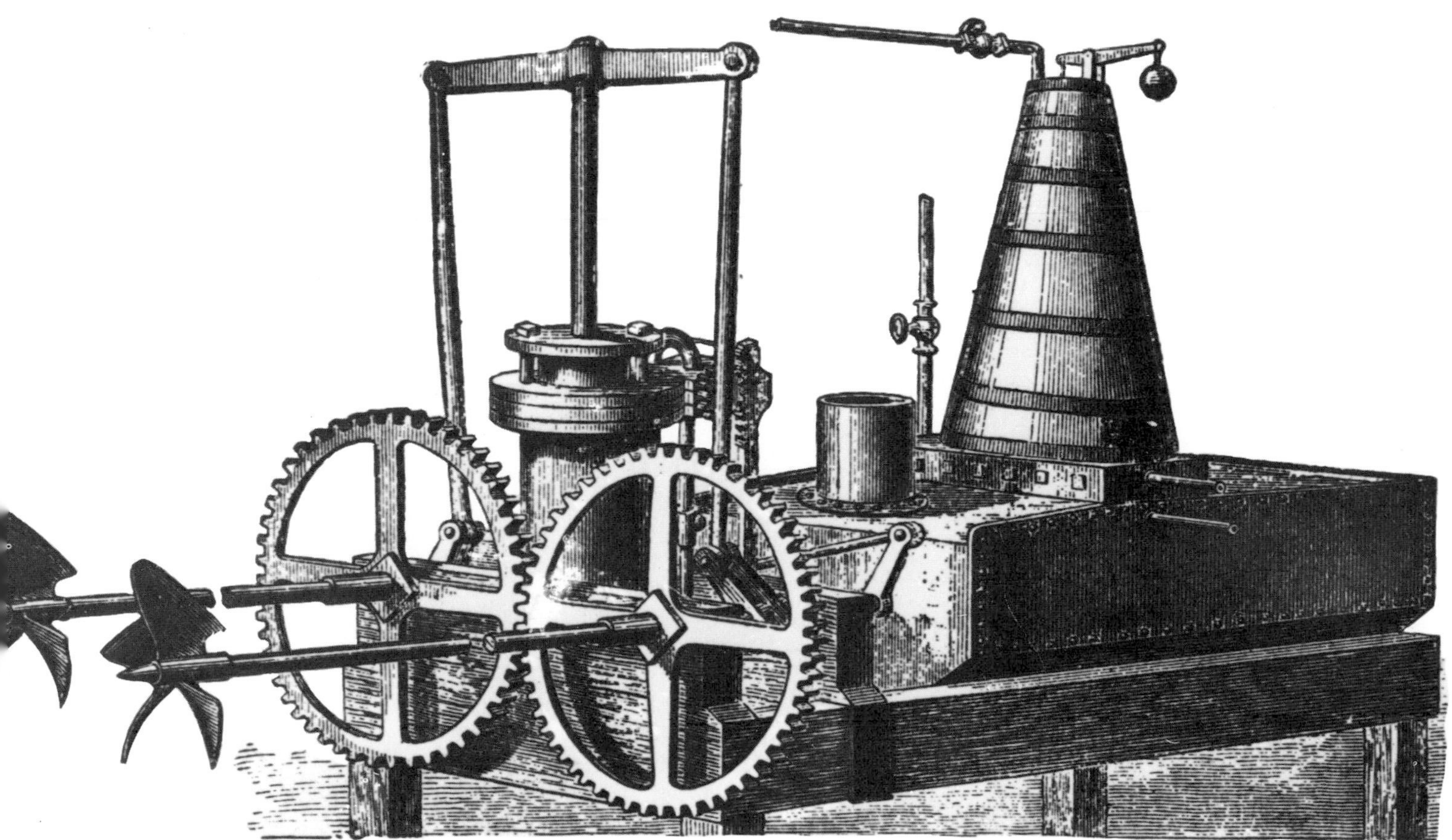

Opposite: Drawing of Robert Fulton as a young man, by John Vanderlyn. Left: Fulton's first submarine, built in France in 1801. This drawing by Fulton shows the boat on the surface and submerged.

By 1793 he was trying to interest a peer, Lord Stanhope, in steam-propelled boats. "My first design," he wrote, "was to imitate the spring in the tail of a Salmon: for this purpose I supposed a large bow to be wound up by the steam engine and the collected force attached to the end of the paddle."

Later he was somewhat more practical with various inventions having to do with canals, a major interest in prerailroad England. One he patented was a means of raising boats from one level of a canal to a higher one without using locks. It involved putting flanged wheels on the boats and hoisting them on tracks to the next level. The weight of descending boats provided the power. Another invention was a means of carrying canals across bridges. This consisted of big cast-iron aqueducts to be cast in sections in sand molds near the canals under construction.

Shortly thereafter he proposed a mechanical excavator for canals, which, had he used steam to operate its mechanism, would have been the world's first steam shovel. Fulton now became excited by an idea he had for building what he called "Creative Canals" which would cover the countryside of England and America. To promote this grandiose scheme of small canals he wrote a "Treatise on the Improvement of Canal Transportation." This not only gave him a modicum of professional standing as an engineer, it also helped him pay off a few of his debts. For he had lived for years by borrowing and scrounging.

Fulton must have been an amazingly persuasive young man. He wrote a friend in 1797 about "having sold one fourth of my canal prospects for £1,500 to a gentleman of large fortune who is going to reside in New York. . . . it is stipulated . . . that I should go to Paris and obtain patents for the small canal system." In France he made friends with Joel Barlow, the American ambassador, and for seven years he lived with Barlow and his wife like a member of the family.

Although he pulled every wire he could to advance his "small canal" ideas, including even Napoleon in his attempts at reaching the great, Fulton soon gave up the idea. He had something new: torpedoes and submarines!

During the uprising in the American colonies, David Bushnell, living in what is now Westbrook, Connecticut, designed a tiny one-man submarine, *The American Turtle.* It launched an unsuccessful attack on HMS *Eagle,* sixty-four guns, in New York harbor, and tried to sink HMS *Cerberus* as she lay off New London but blew up a schooner anchored astern of her instead.

In 1797, then, the idea of a submarine was not exactly new. But this did not deter Fulton from presenting his plans for the submersible *Nautilus* to the French Directory as a new invention. For it was in reality an entirely new approach to the problem of an undersea vessel of war.

Fulton's *Nautilus* was surprisingly like a modern submarine. It had ballast tanks filled and emptied by a pump, a conning tower, a screw propeller (hand-operated), and diving planes. A fan-like sail which folded into the deck and was covered by twin folding fairings when submerged, was available for propulsion on the surface.

Fulton actually succeeded in having his submarine built. In a letter to the commissioners appointed by Napoleon "to promote the invention of Submarine Navigation," he describes one of his many trials: "On the 3rd of Thermidor [July 22, 1801] I commenced my experiments by plunging to the depth of 5, then 10 . . . and to 25 feet, but not to a greater depth than 25 feet as I did not conceive the machine sufficiently strong to bear the pressure of a greater column of water. At this depth I remained one hour with my three companions and two candles burning without experiencing the least inconvenience."

A few days later, Fulton tried his submarine under sail, then wrote: "I lowered the mast and sails and commenced the operation of Plunging. I then placed two men at the engine which gives the Rectilinear motion, and one at the helm, while I governed the machine which keeps her balanced two ways. With the bathometer before me and with one hand I found I could keep her at any depth I thought proper. The men then commenced their movement and continued about 7 minutes when mounting to the surface I found we had gained 400 metres."

Fulton then went on to describe how he maneuvered the *Nautilus* in every possible direction. He planned a brass

Preceding pages: The *North River Steamboat,* **later called the** *Clermont.* **Right: Fulton's drawing of the** *Clermont's* **propulsion system. The engine was built by Boulton and Watt in England. Opposite, top: The** *Clermont* **comes in for a landing at Cornwall-on-Hudson. Bottom: 1813 drawing by Fulton of his floating fortress** *Fulton the First,* **or** *Demologos.* **From the top, the drawing shows a transverse section, an overhead view of the gun deck, and a side view of the vessel.**

sphere to contain oxygen or compressed air which would allow the crew to remain submerged for six hours. He also described a test with a bomb: "They are constructed of copper . . . to contain 20 to 200 pounds of powder. Each bomb is arranged with a Gunlock in such a manner that if it strikes a vessel . . . the explosion will take place. . . . Admiral Villaret ordered a small Sloop of about 40 feet long to be anchored . . . with a bomb containing about 20 pounds of powder I advanced to within about 200 metres; then taking my direction so as to pass near the Sloop, I struck her with the bomb in my passage. The explosion took place and the sloop was torn into atoms."

Earlier, Fulton had actually tried going after the British fleet, a feat of great daring considering the fragility of his little craft. But the British, remembering Bushnell's *American Turtle,* kept strictly away from the *Nautilus* under orders not to fight. Tradition has it, however, that Fulton came breathtakingly close to sticking one of his torpedoes into the bottom of a British seventy-four.

Still, Fulton failed to get the French admiralty to adopt his submarine, mostly because the mossback admirals thought it a most unchivalrous kind of weapon, but partly because Napoleon thought it impractical. Later, after he was already involved with steamboats, Fulton carried on torpedo and submarine experiments for the British admiralty, but its interest seemed chiefly in suppressing such dangerous and ungentlemanly threats to Britannia's rule of the waves.

It was while Fulton was experimenting with submarines that Livingston—now Chancellor Robert R. Livingston, Minister Plenipotentiary of the United States to France—met him and started talking steamboat to him.

Livingston was aware of Fulton's engineering background and was especially impressed by his work on the submarine. Compared to the urbane and professional Fulton, Stevens, way off in Hoboken, seemed a mere putterer. He *had* been taking an unconscionably long time arriving at a viable steamboat.

Fulton listened hard to Livingston, who soon convinced him of the need for steamboats on roadless America's long rivers—the Hudson, the Mississippi, the Ohio, and the rest. Fulton lost little time in starting his experiments, especially since Livingston not only offered solid financial backing, but also owned Fitch's old monopoly for navigating the Hudson. By the spring of 1803, Fulton had his steamboat afloat on the Seine. It was 66 feet long and had an eight-foot beam. He used a borrowed single-action engine with a small brass cylinder of two-inch bore and 22-inch stroke. He had at first used a remarkably advanced flash-steam boiler of his own design and placed the cylinder within the boiler to keep it as hot as possible. But the metals of 1803 were not able to withstand high temperatures and the boiler was changed to a more conventional type.

But Fulton had bad luck. One night, while his boat lay in the Seine, a violent storm started it rolling and plunging at its mooring. Too lightly built to take such strains with heavy machinery aboard, it broke up. The engine went through the bottom to the riverbed.

Fulton didn't give up. He retrieved the machinery, put it into a slightly bigger 74.6-foot hull. A Paris newspaper described its trial: "There has been seen at the end of the Quay Chaillot a boat of curious appearance, equipped with two large wheels mounted on an axle, like a cart, while behind these wheels was a kind of large stove with a pipe. . . . Assisted by three persons only, he put his boat in motion with two other boats in tow behind it, and for an hour and a half he afforded the curious spectacle of a boat moved by wheels like a cart, these wheels being provided with paddles or flat plates and being moved by a fire engine.

"In following it along the quay the speed against the current of the Seine appeared to us about that of a rapid pedestrian, that is about 2400 *toises* [2.9 miles] per hour."

Fulton was now ready to build the steam-driven vessel which would conquer the Hudson. But before returning to the United States he spent that fruitless time dealing with the admiralty as it stalled and bumbled over his submarine and its torpedoes. But Fulton's time wasn't entirely wasted. It was then that he became acquainted with Symington's *Charlotte Dundas.* It was then, too—in 1804—that he paid a visit to Boulton and Watt at Soho, where in his usual charming manner he made friends with the great Matthew Boulton him-

self; Watt had already retired.

While in France, Fulton had tried to order a Watt steam engine for his new boat. But under British law, no mechanical devices or, indeed, any mechanics might leave England. But by 1804 the restrictive law was no longer in force and Fulton was able to order his engine. This, with a 24-inch bore cylinder and a 48-inch stroke, with a condenser and "Working gear compleat with Brackets" and various spanners, rods, pipes, nozzles, pumps, and "two boxes of cement," cost £380.

Arriving back in New York after nineteen years abroad, Fulton immediately got to work, not only on the steamboat but also on further torpedo experiments for the U.S. Government. These came to an end after a Navy commodore slyly rigged the world's first antitorpedo nettings on the sloop-of-war *Argus* which Fulton was to attack with his new weapon.

By March, 1807, the *North River Steamboat* was a-building in Charles Browne's yard at Corlear's Hook. The hull's ugly shape was based on water-resistance tests made by Colonel Mark Beaufroy in Greenland Dock in London ten years earlier. She looked like a long (146-foot), narrow (12-foot), rectangular, flat-bottomed barge with crudely pointed ends. She was really too narrow for safety.

By May her engine was aboard. In August Fulton tested her and experimented with various paddle widths. She was not quite finished; the paddle wheels were not boxed in and her engine was uncovered. Someone said she looked like "a backwoods sawmill mounted on a scow and set on fire."

But on August 17, 1807, with forty of Fulton's friends and relatives aboard, the *North River* left her New York dock for a trial run.

Fulton wrote his friend Barlow: "My steam boat voyage to Albany and back has turned out rather more favourably than I had calculated. The distance from New York to Albany is one hundred and fifty miles. I ran it up in thirty-two hours and down in thirty. I had a light breeze against me the whole way both going and coming and the voyage has been performed wholly by the power of the steam engine. I overtook many sloops and schooners beating to windward and parted

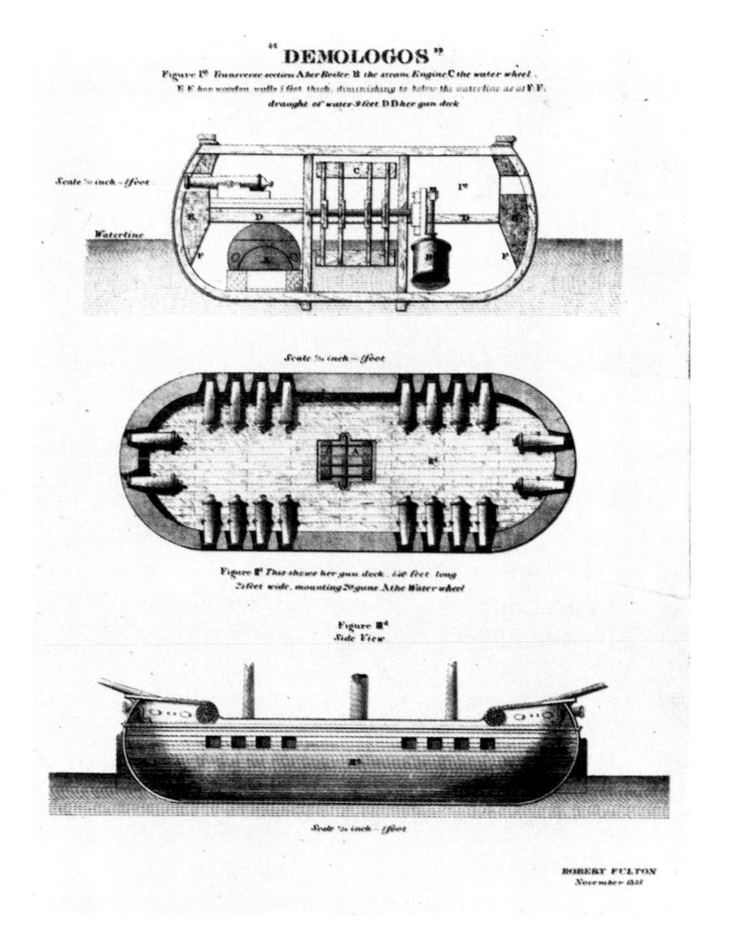

1. *Great Eastern* **laying the Atlantic cable.**
2. Isambard Kingdom Brunel, designer of *Great Eastern,* **was the foremost engineering genius of the Victorian era.**
3. *Great Eastern* **resting on the mudbank at Millwall-on-Thames.**
4. London Science Museum model of the *Mauretania,* **built in 1906, when steamers were the queens of the North Atlantic.**

2

1

3

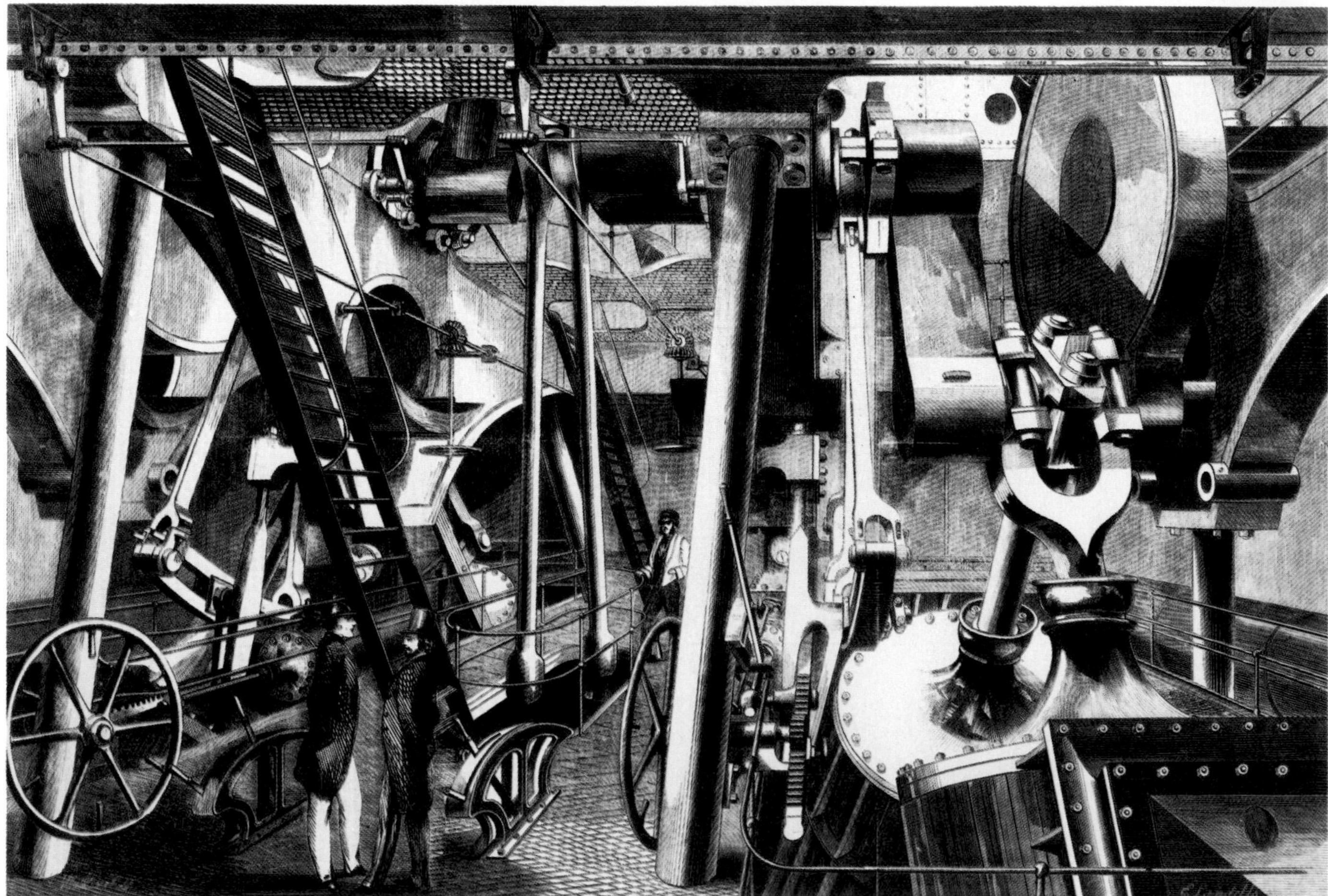

Opposite, top: *Great Eastern* **arriving in Liverpool. Bottom:** *Great Eastern's* **two huge reciprocating engines drove both paddle wheels and screw propellers. The cylinders of the Boulton and Watt engines were seven feet in diameter.**

with them as if they had been at anchor."

In September the Albany *Gazette* ran an advertisement:

> *THE NORTH RIVER STEAMBOAT*
> will leave Pauler's Hook Ferry on Friday the 4th of September at six in the morning and Arrive at Albany at 6 in the afternoon. Provisions, good berths and accommodations are provided.

The age of steam transport had at last truly begun.

In spite of small mishaps—a sailing packet skipper purposely ramming a paddle wheel (thenceforth the wheels were enclosed), and a broken cast-iron axle shaft—the *North River* continued its runs until the Hudson froze.

Fulton was able to write Livingston: "After all accidents and delays our boat has cleared 5 per cent. . . . A boat which will cost us $15,000 will . . . provide us $10,000 a year."

A new, stronger hull was built for the *North River,* the New York legislature extended the monopoly for five years longer, and a second boat, the *Car of Neptune,* was built in 1809. It was now that the name of the *North River* was changed to *Clermont,* after the name of Livingston's estate.

Fulton went on to build ferries, other bigger steamboats, and in 1811 the *New Orleans,* the first steamboat on the Mississippi. For the war of 1812 he built a monstrous, 300-foot, steam-propelled floating fortress, the *Demologos* or *Fulton the First.* But the war ended before she was commissioned and the *Demologos* became a depot vessel at the Brooklyn Navy Yard, where she accidentally blew up in 1829.

Fulton died in 1815, only fifty years old.

In 1818 the American ship *Savannah,* the first vessel to use steam power in addition to its sails, crossed the Atlantic. In 1837 the British *Sirius* was the first to cross entirely under steam.

By mid-century steam was king of the seas for mail and passengers. It was then that that wondrous failure, "the great iron ship," "the Floating City," "the Crystal Palace of the Sea," or more prosaically, the *Great Eastern,* was built. Isambard Kingdom Brunel, one of the great engineers of all time, and John Scott Russell designed her and saw her keel laid in the yard of Messrs. John Scott Russell & Company, Millwall.

No ship equal in size to the *Great Eastern* was launched for another forty years. Six hundred and eighty feet long and with an 82.7-foot beam, she had quarters for 800 first-class passengers, 2,000 second class, and 1,200 third class. Her grand saloon, all gilt and white, was 62 feet long. Her staterooms offered twice the space of those in the most luxurious Cunarders. Concealed in the stateroom settees were baths with hot fresh water and cold salt water.

Her paddle wheels were 56 feet in diameter, her screw propeller 24 feet in diameter. Two sets of engines drove both independently. But the *Great Eastern* was, by our standards, underpowered. A 1,000-hp, four-cylinder engine drove the paddle wheels, and 1,600-hp, four-cylinder engines the screw. The cylinders of the Boulton and Watt screw engines were no less than 84 inches in diameter! By comparison, the *Queen Mary* of 1934 had 160,000 hp. The top speed of the *Great Eastern* wasn't much over sixteen knots, the *Mary's* about thirty knots.

Great Eastern's troubles started before she was even afloat. She was to be launched sideways into the Thames late in 1857. But she balked on the way. It became necessary to use hydraulic jacks which moved her inch by inch, and she wasn't floated until the end of the following January. Launching costs and Brunel's problems with the company directors forced the company into bankruptcy. The *Great Eastern* then lay at anchor in the Thames for twenty months. Brunel and some of the original directors bought her and fitted her out for sea, but on her trial run she suffered an explosion in the water jacket of a funnel. She was taken to Holyhead, Wales, where she was opened to sightseers. A great storm almost wrecked her.

Another board of directors, in spite of the fact that the ship had been designed for long passages to India, sent her on the short run to New York in 1860 with only thirty-eight paying passengers aboard. In America paying sightseers vandalized her and covered her decks with filth. On another trip to Quebec forty passengers, 2,500 soldiers, and 200 horses made the crossing.

Preceding pages: Prior Park, 1731. The first railways had wooden tracks and lacked locomotives. Here, cars laden with stone are moved by gravity toward barges on the River Avon.
1. Richard Trevithick. 2. Trevithick's model locomotive of 1798.
3. Oliver Evans' ***Orukter Amphibolus.***
4. Trevithick's road-going steam carriage of 1801.

1

2

Later she ran aground and holed her bottom near Montauk Point, Long Island. *Great Eastern's* unrelieved bad luck continued until 1865, when she was chartered to lay the Atlantic cable, a job successfully completed.

When she went to the scrap yard in Liverpool in 1888, she even gave the scrappers trouble. It took three years to cut her up.

For a hundred years ocean-going steamers grew ever bigger, faster, and more luxurious. They carried the mails and the rich. But mostly they carried the millions of poor immigrants who made the United States what it is today.

The day of the giant passenger steamship is almost over. Those left, like Cunard's *QE II,* lose fortunes on their transatlantic runs, which they recoup in part with excursions carrying winter trippers to tropical resorts.

Steam on the Rails

It is the year 1798. In Richard Trevithick's house in the Cornish town of Camborne a group of friends sits around the table in a long, low-ceilinged room. Dinner has been dealt with and the big oak table has been cleared except for a few wine bottles and pewter beakers.

Trevithick is kneeling in front of the wide fireplace. He removes a kettle from the crane and pours hot water into a foot-long iron cylinder on wheels which rests on the hearth. He then takes from the fire a red-hot iron—the kind used at the time for heating tea urns—and inserts it into a tube in the little device. He waits a few moments, then moves a tiny valve. The mechanism starts puffing as a rod slowly emerges from one end and moves other rods attached to the wheels, which start turning. Trevithick lifts the hot little machine with a folded napkin and quickly sets it down, pointing toward the far end of the room. Suddenly it seems alive. Puffing away, its rods a blur, it flies along the floor until it hits the wall with a bang.

Trevithick's friends laugh and cheer. (There has been much wine at dinner.) Trevithick, much pleased, and also in a bit of a glow, retrieves the iron model and sets it on the table where it skitters, puffing and snorting, the length of the table until stopped, wheels still spinning, by Trevithick who has run alongside to prevent its falling over the edge.

Trevithick had just shown his friends the prototype of his steam locomotive. In six years his full-sized locomotive will run on rails. Ironically, however, George Stephenson, not Richard Trevithick, will be known as "the father of railways."

Carts had been run on rails for centuries. Self-propelled steam vehicles had been conceived by earlier inventors, and in some cases actually operated. John Fitch, William Murdock, Nathan Read, and Nicholas Cugnot, among others, had had the idea of using steam power for transport. Both Murdock and Fitch built models. Murdock actually ran his in 1784, and Cugnot built a full-sized machine which is said to have operated as early as 1769. None of these was really a precursor of the railway locomotive, although Fitch had foresightedly fitted flanged wheels to his model of 1796.

The success of rail- or road-going vehicles depended on finding a way to utilize high-pressure steam. Newcomen's and Watt's engines had used steam only a few pounds above atmospheric pressure. To realize the kind of energy needed to propel a vehicle at a useful speed, a Watt engine would have had to be as big as a house. Portable power from a Watt engine was an impossibility. Two men who understood this were the American Oliver Evans and the Cornishman Richard Trevithick.

Evans never actually built a railway locomotive. He did, however, build the famed *Orukter Amphibolus* in 1804, on commission from the Philadelphia Board of Health for dredging and cleaning docks. Evans had built his machine sixteen miles up river and a mile and a half from the water. Since it reached its dock-cleaning position under its own power, it turned out to be not only a steam-operated dredge, but also a self-propelled land vehicle—as well as a steamboat run by a paddle wheel. Although painfully slow, it is obvious that its twenty-ton weight could only have been moved by a fairly efficient little engine—a high-pressure engine. Evans

3

4

had used pressure remarkably high for 1804: forty pounds per square inch.

But Richard Trevithick had been far ahead of Evans. Trevithick was brought up in steam. His father, Richard senior, had spent much of his life as manager of a tin mine in Cornwall and was closely involved with the Newcomen and Watt engines used for pumping the mines.

As a boy, young Trevithick (born in 1771) was wild about the steam engines surrounding him, and neglected his schooling to fiddle with them. By the time he was nineteen he was a steam engineer at several mines.

The Cornish miners had been greatly benefited by the Watt engine. Although it reduced the cost of fuel, since it was incomparably more efficient than their old Newcomen engines, they were outraged at having to pay the Boulton & Watt Company royalties for its use. The mine operators were particularly infuriated by Watt's monopolistic patent and continually sought ways to get around it.

Trevithick was one of the ingenious mechanics who designed engines in hopes of circumventing the Watt patent. In fact, he became such a nuisance to Boulton & Watt that they seriously considered employing him—one way of getting him out of their hair. But B & W offered him too low a salary. Further, William Murdock, Boulton & Watt's very important employee, took a dim view of associating with Trevithick.

Trevithick wasn't really the kind of young man who'd fit in with a solid and conservative firm. He was a whirlwind, six feet two inches tall, handsome, exuberant, and rather stronger than some bulls. And he was at times prone to show off that strength in pubs. His tussle with a certain Captain Hodge at a mine owner's dinner was well remembered. Hodge was as big as Trevithick, but found himself grabbed around the waist, turned upside down, and heaved upward until his boot prints were stamped on the ceiling of the inn. Another little trick of Trevithick's was writing his name on a ceiling beam with a sixty-pound weight hanging from his thumb. Still another of his stunts was lifting an eleven-hundred-pound mandrel in a smithy. Hundreds of people used to come watch that one. And he could out-wrestle any man in town.

By 1797, when he was twenty-six, Trevithick had enough experience to realize that huge engines, with their monstrous cylinders, air pumps, and condensers, and their ponderous, slowly rocking beams, were passé. What was needed was something smaller and more muscular—an engine that could be built cheaper, go anywhere, and do any kind of work.

Trevithick's first road-going vehicle using high-pressure steam was tried out on Christmas Eve, 1801, in Camborne. Trevithick, instead of making his first attempt on a level road, immediately tried to climb a hill. The boiler ran out of steam and the engine stopped. A few days later, after Trevithick had made some adjustments, the carriage ran successfully for a time, giving some of Trevithick's friends their first exciting sensation of mechanical locomotion. But the machine balked again, and in disgust Trevithick, Cousin Andrew Vivian, who had helped in its construction, and a few friends pushed it into a barn. One, a Mr. Giddy, described what happened next: "The parties adjourned to the Hotel and comforted their Hearts with a Roast Goose & proper drinks, when forgetfull of the Engine, its Water boiled away, the iron became red hot, and nothing that was combustible remained either of the Engine or the house."

Trevithick's second carriage was assembled in London two years later. It was a three-wheeler with eight-foot driving wheels and boiler and engine between the wheels. It held eight or ten people. The engine was wonderfully compact for its time, "about the size of an orchestra drum," one observer wrote. The 5½-inch bore, two-foot six-inch stroke cylinder was inside the small boiler and did fifty strokes per minute at thirty pounds per square inch pressure.

On its first trial it ran from Leather Lane, where it had been built, "through Liquorpond Street into Gray's Inn Road by Lord's Cricket Ground to Paddington and Islington and back." Later trips were made in Tottenham Court Road and in Euston Square. Speed was eight or nine miles an hour. But no one, including the newspapers, seems to have taken even slight notice of this remarkable feat. Trevithick gave up on road carriages.

He then turned his active mind to using the tracks

Preceding pages: Model of Richard Trevithick's 1804 locomotive.
Right: Trevithick's drawing of his 1804 locomotive.
Opposite, top: Ticket for a ride on *Catch Me Who Can,* which ran in 1808 on a circular track near what is now Euston Station, London.
Bottom: Drawing of the fenced-in track by famous cartoonist Rowlandson.

on which horse-drawn wagons were run by mine operators. One of these operators, a Mr. Hornfray, for whom Trevithick had built stationary engines, was a sporting type. He made a bet with one Anthony Hill, who owned the Plymouth Ironworks, that he could haul ten tons of iron by means of a steam engine on the tracks from Penydaron to Abercynon, a run of nearly ten miles. Hill bet even money it couldn't be done—a thumping 500 guineas.

In February, 1804, Trevithick had his locomotive—the world's first locomotive—ready and running. On the twenty-second he wrote:

"Yesterday we proceeded on our journey with the engine; we carry'd ten tons of Iron, five waggons, and 70 Men riding on them the whole of the journey. Its above 9 miles which we perform'd in 4 hours & 5 Mints, but we had to cut down some trees and remove some Large rocks out of road. The engine, while working, went nearly 5 miles pr hour . . . the gentleman that bet five hund.d guineas against it, rid the whole of the journey with us and is satisfyde that he have lost the bet. We shall continue to work on the road, and shall take 40 tons the next journey. The publick untill now call'd mee a schemeing fellow but now their tone is much alter'd."

Besides using high pressure, there was another important Trevithick innovation. The exhaust steam, which in Trevithick's earlier high-pressure engines had been released into the atmosphere without going into a condenser, as in a Watt engine, now was directed up the chimney, increasing the blast and making the fire hotter. Every successful locomotive of the future used such a forced draft.

In spite of Trevithick's highly successful demonstration there was no rush to build railroads. The problem was tracks. Trevithick's five-ton locomotive was too heavy and broke the early cast-iron rails. Nonetheless, he had proved it could be done and, typically, he went off in a dozen other directions. He built rolling-mill engines, dredges, a weird steam machine which, mounted on a barge and using a gargantuan chisel, broke underwater rocks, and he even became involved with an idea for a steamboat to tow fire ships into Boulogne harbor in order to destroy Napoleon's invasion fleet, then poised to invade England. Once when transporting an engine by barge to a cotton factory, he hooked it up to a couple of roughly thrown-together paddle wheels. "It wod go in still water abt 7 Miles p. Hour," he wrote.

In 1805 Trevithick became deeply involved with building a tunnel under the Thames. This failed after he had driven a thousand of the twelve hundred feet of driftway, or trial tunnel, he had to go (Brunel had not yet invented the tunneling shield) and Trevithick found himself, for once, with time on his hands.

He had one last affair with locomotives. The *London Times* of July 8, 1808, said: "We are credibly informed that there is a Steam Engine now preparing to run against any mare, horse, or gelding that may be produced at the next October Meeting at Newmarket; the wagers at present are stated to be £10,000; the engine is the favourite. ... *Trevithick,* the proprietor and patentee of this engine, has been applied to by several distinguished personages to exhibit this engine to the public . . . its greatest speed will be 20 miles in one hour."

Trevithick did not run his engine on a race course "against any mare, horse or gelding," but he did run his portable steam engine, which he named *Catch Me Who Can,* on a circular track enclosed by a fence, near what is now Euston Station. He charged five shillings to watch or, for those without fear, to ride the snorting ten-ton beast. The locomotive ran as advertised and reached a speed of 12 mph (Trevithick said it would go 20 mph on a *straight* track), but there were too few ticket-buyers to pay expenses. Anyhow, Trevithick had again proved his point and thereupon gave up locomotives for good.

He then turned to a dizzying number of engineering projects: steam lighters for use in the port of London, steam threshing engines, cultivators, factory engines, new types of boilers, tugboat engines, screw propellers. He went through bankruptcy.

In 1813 he became involved in a scheme for pumping out the silver mines of Peru. He built the engines for this venture, which involved transporting them over the wild and narrow mule paths of the Andes. In 1816, he himself set sail for Peru. He spent eleven years in South America. Robbed, cheated by the Spaniards, mixed up in revolutions, he at last

My ride with
Trevithick in
the year 1808 in
an open Carriage
propelled by the
Steam Engine of
which the Enclosed
is a print, ~~it~~ took
place, then a Waste
piece now Torrington
Square.

TREVITHICKS,
PORTABLE STEAM ENGINE.
Catch me who can.
Mechanical Power Subduing
Animal Speed.
p 366
No 9

Right: Blenkinsop's 1812 locomotive was geared to the track. Below: 1813 locomotive *Puffing Billy,* now in London's Science Museum. Opposite: Reproduction of *Locomotion,* built for the 150th anniversary of the Stockton & Darlington Railway, steams along a modern track.

Opposite, top left: George Stephenson, "Father of the Railway." Top right: Robert Stephenson. Bottom: Painting of the Stephenson family, showing George Stephenson (center), his son Robert, and a man who may be his father. George Stephenson holds the miner's lamp he invented.

returned to England after a desperate journey across Central America. He landed penniless with "the clothes he stood in, a gold watch, a drawing compass, and a pair of silver spurs."

Trevithick soon busied himself with other ventures: ideas for quickly reloading cannon on warships, methods for draining the polders of Holland, new types of steam engines, steam turbines, and boilers, even a "warming machine" for houses. In 1832 he designed a gigantic, screwed-together, cast-iron column a thousand feet high to commemorate the passing of the Reform Bill. But in 1833, before any of these ideas came to fruition, he died. He was sixty-seven. Others had already made a success of the steam railroad.

While Trevithick was tilting with windmills in South America, the industrial climate of England had changed greatly. The canal builders, inventions in the textile industry, the stationary engines of Boulton & Watt, and the Napoleonic wars had caused an industrial explosion. The time was ripe for railroads to transport the raw materials, the finished goods, and the burgeoning populations of the Industrial Revolution.

Trackage upon which cars were pulled by horses had long existed. And, of course, Trevithick had proved that a "portable steam engine," a locomotive, could be run similarly on such tracks. There was a weight problem; Trevithick's heavy locomotive broke the fragile tracks, yet it was feared that lighter locomotives with smooth wheels would not have enough traction. To make sure that it wouldn't just stand there spinning its wheels when the throttle was opened, the first locomotive built in 1812, after Trevithick had turned to other pursuits, had a geared driving wheel which engaged a toothed track. The "geared to the track" idea was John Blenkinsop's. The locomotive was built by Matthew Murray to operate on a wooden track to haul coal. Within a year four such Blenkinsop/Murray locomotives were at work.

The adhesion problem was attacked in other ways. William Brunton's locomotive had legs and feet actuated by an antic system of levers. William Chapman designed a locomotive that hauled itself along a chain cable laid between the rails.

Another famous early locomotive builder was William Hedley, who came forth with *Puffing Billy* in 1813. Amazingly, it remained in more or less continuous operation until 1862, when it was acquired by London's Science Museum, where you can see it today.

But inventing locomotives was not enough. England and the world needed railroad systems. Two men who saw this and who should be given credit for pioneering the railway as a viable means of transporting freight and passengers were George Stephenson and his son Robert.

George Stephenson, born in 1781, was in charge of a pumping engine at a Northumberland coal mine by the time he was seventeen. Thereafter he was a brakesman in charge of the winding engine which raised coal to the surface and also brought the miners to and from the coal face.

Ambitious, shrewd, and very tough, he had practically no formal education. Recognizing this, he decided that his son Robert, born in 1803, would be the reader, writer, and arithmetician of the family. Almost from infancy, poor Robert had little respite from books and tutors and schools. Nor did Stephenson neglect his son's mechanical education. He enlisted him as a helper in repairing clocks, a sideline Stephenson *père* had taken up to make some extra money. They even made a sundial together for which Robert had to calculate the exact latitude and longitude.

While still a brakesman, George Stephenson showed remarkable mechanical abilities. In 1811 a newly installed Newcomen-type pumping engine designed by John Smeaton failed to pump. Respected steam-engine boffins were baffled by it. Brash young Stephenson claimed he could make it work. The mine owners, desperate, allowed the lowly brakesman to have a go. He revised the water-injection system and raised the steam pressure—unheard-of on a Newcomen engine. To the alarm of onlookers it almost rocked the engine house off its foundation when it started, but in two days the mine was pumped dry. Soon he was no longer a brakesman, but an engine-wright who traveled from mine to mine. Within a few years he was building engines by the dozen, including the great Friars' Goose pumping engine. This giant had a six-foot bore and a nine-foot stroke and pumped a thousand gallons a minute from a depth of fifty fathoms.

The Liverpool & Manchester Railway, opened for operation in September, 1830.
1. A locomotive taking on water at Parkside.
2. The tunnel, lit by gas.
3. A *Northumbrian*-type locomotive and rolling stock.
4. The "Railway Office" in Liverpool.

1

2

3

4

1

2

Perhaps the greatest spur to Stephenson's fame, however, was his invention of a miner's safety lamp in 1815, before Sir Humphrey Davy invented his. Sir Humphrey still gets the credit, but Stephenson's "Geordie" lamp was for years preferred by miners in the Killingworth area.

Railway tracks were an old story to George Stephenson. The Wylam Wagonway had run past his birthplace, and he had seen locomotives running on such lines, as well as horse-drawn coal cars.

He had long been eager to try his hand at locomotive building and his chance came when Sir Thomas Liddell ordered him to oversee the construction of an engine to run on the Killingworth wagonway. That first locomotive was named Blucher and it made its first slow rumbling run past Stephenson's cottage at West Moor in July, 1814.

The Blucher was not a revolutionary design. It did, however, have an important new feature—flanged wheels. It also had a serious failing: a tendency to run out of steam. Stephenson tried various methods to improve Blucher's steaming qualities: turning the exhaust up the chimney as Trevithick had done, and increasing the diameter of the boiler flue. But Blucher still sometimes stalled embarrassingly.

James Stephenson, George's older brother, was the Blucher's first driver, and his fireman was James Ward, who described one of the times the Blucher ran out of power. It was hauling a twelve-car train of coal weighing thirty-six tons and was slowly crossing a road near James's house. Puffing and spinning its wheels and making almost no progress, the Blucher was blocking the road. Through a window James could see his wife doing her housework. "Come on, Jinnie," he bellowed, "put your shoulder to her." Jinnie, a most buxom lady, downed her broom and pushed with enough effect to get the Blucher going again. More than that, Jinnie is said to have risen daily at four in the morning to light the Blucher's fire.

George Stephenson improved both locomotives and the tracks they ran on. In his day tracks still were mounted on stone blocks. The cast-iron rails were butt-ended, and if a block were canted one rail would be raised higher than the other at the joint, thus causing derailments. Stephenson devised lap joints and rail seatings to prevent this. A "steam cushion" device for springing locomotives was another of his ingenious (but not too successful) notions.

But a few feeble locomotives hauling coal out of mines for short distances did not constitute railways. Stephenson had a larger view, and in 1821 he got his chance. By then he was a fairly well-known and respected engineer, even though he did speak in a barely intelligible Northumberland brogue and was obviously of lower-class origin. He was chosen to survey and estimate the costs of a twenty-four-mile railway line between the mines of Stockton and the River Tee, where the coal could be loaded on barges.

George Stephenson, his son Robert—freed at last from the mine in which he worked—and a few assistants surveyed the right of way. George designed the bridges, including the first iron bridge ever used on a railroad.

Earlier proposals had suggested digging a canal and building tramways for horse-drawn cars, but now, after years of indecision, steam locomotives were the choice. Wrought-iron rails replaced the older cast iron, and the gauge was four feet eight inches, a span which went back to the Roman chariot. (Most railroads since have added a half inch to the track width. No one seems to know why.)

The rails were supported part of the way by stone supports, the rest by wooden blocks cut from the timbers of superannuated naval vessels. Since horses were expected to do some of the hauling, full-width ties that would interfere with their stride were not permitted.

Stephenson estimated that the Stockton & Darlington line would cost £74,300 and in May, 1822, the first rail was laid to the music of bands and the roar of cannon. By September, 1825, the world's first public railroad was ready for traffic. And a new locomotive, the *Locomotion,* designed by George Stephenson and built by a firm organized by him and twenty-year-old Robert, Robert Stephenson & Co., was ready to run on it.

The *Locomotion* was no remarkable advance on earlier Stephenson engines, but it performed well that day, preceded by a horseman and hauling coal cars filled with an overload of six-hundred-odd roistering passengers.

Running a locomotive in 1825 wasn't a simple task.

1. The famous locomotive *Rocket*, winner of the Rainhill Trials.
2. Braithwaite and Ericsson's *Novelty*, built by John Ericsson, who later built the Union navy's *Monitor*.
3. The locomotive *Northumbrian*.

3

Locomotion was brought to the track on a wagon. Even the drill of lighting her fire that day makes us smile. Workmen first filled her boiler and threw chunks of wood into the firebox. But how to light her fire? No one carried matches in those days. Johnny Taylor was sent off to bring back a lighted candle lantern, but he took forever. Had he stopped for a dram on the way? A bystander taking out his "pipe glass," a burning glass, to light his pipe notices a spare bit of oakum packing for the water-feed pump which has been sent along in case the pump needs repacking. He focuses the sun on the oakum. It flames and he throws it through the fire door. Soon, the air above *Locomotion's* boiler starts to shimmer with heat.

Even before the Stockton & Darlington was running, George Stephenson was involved in another, much grander operation, the real ancestor of fast passenger and freight railways: the thirty-mile Liverpool & Manchester. (Robert Stephenson had little part in this project. He left for South America for reasons not too dissimilar from those which had enmeshed Richard Trevithick.)

The Stockton & Darlington, somehow, had not encountered opposition. The Liverpool & Manchester was a different matter. The vested interests—big landowners, canal and turnpike companies, and stagecoach lines—were aroused. Tavern keepers and wagoners fearful of competition and simple folk fearful of innovation worked themselves into fevers of excitement. Survey parties were stoned, screamed at, and ambushed by landowners' men armed with shotguns. A chainman was impaled on a pitchfork.

Of the unruly estate owners, Stephenson wrote: "Their ground is blockaded on every side to prevent us getting on with the survey. Bradshaw fires guns through his ground in the course of the night to prevent the surveyors coming in the dark. . . . Lord Sefton says he will have 100 men to stop us."

Once work actually started, frightful engineering problems were encountered. A great morass, Chat Moss, had to be crossed. Engineers and laborers could only venture into the swamp by tying boards to their feet. The banks of ditches to drain the area of the right of way oozed away as fast as they were dug. Stephenson finally conquered Chat Moss by building a miles-long floating raft of brush and branches on which he raised his embankment. Thousands of tons of earth and stone were moved—in those days before bulldozers—by men with shovels and horses dragging carts.

There was still some doubt as to whether locomotives would be the sole motive power on the L & M. Horses were strong contenders, as were fixed engines which would wind the cars along by ropes. In fact, a majority of the Railway Board originally leaned toward fixed engines set at intervals along the line.

Another group, steam enthusiasts eager to prove that such motive power was eminently practical, proposed giving a £500 prize for the best locomotive. The entire board agreed and the great, the famous, the momentous Rainhill Locomotive Trials resulted.

Robert Stephenson had been back from South America for two years by the time the Rainhill trials took place in 1829. During that time he had revitalized his engine works and had built a much improved locomotive, the *Lancashire Witch*.

Stephenson had been much impressed by the builders of steam road carriages, those gaudy precursors of the automobile which had a brief and fairly successful life until killed off by brutal taxation and the attacks of turnpike owners and the railroads. He wrote a friend: "I have been talking a great deal to my father about endeavouring to reduce the size and ugliness of our travelling-engines, by applying the engine [meaning cylinder] either on the side of the boiler or beneath it entirely, somewhat similarly to Gurney's steam coach." (Sir Goldsworthy Gurney was one of the pioneers of steam locomotion on highways.)

The *Lancashire Witch*, the result of Robert Stephenson's hard look at the steam carriages, was a new kind of locomotive, lighter and no longer a stationary beam engine set to propelling itself on tracks. Its outside cylinders were no longer set vertically but at an angle, with their piston rods angled toward the driving wheels, thus eliminating the rods and levers that complicated earlier engines. This arrangement permitted all four wheels to be sprung, which lessened pounding on the fragile tracks and reduced breakage. The *Lancashire Witch* was a success and the prototype

Opposite, top: Model of Frenchman Marc Seguin's 1829 locomotive. Built for use on the St. Étienne-Lyons line, it was the first locomotive to use a multitubular boiler. Impeller fans on the tender acted as bellows to fan the fire.
Bottom: John Stevens' 1826 locomotive.

for the *Rocket,* which went on to win the Rainhill Trials and become, like Fulton's *Clermont* and the Wright *Flyer,* one of the best-remembered transport vehicles of all time.

Many proposals for the Rainhill Trials were fascinatingly eccentric. Like the weird notions the late nineteenth century had about how to propel automobiles— "by means of pendulums," "by the weight of the passengers," and similar fantasies—everything conceivable as a propulsive force was suggested, from perpetual motion to columns of mercury.

But the rules precluded such peculiarities. The engines were to weigh less than four and a half tons if mounted on four wheels, under six tons if on six—including a boilerful of water. A six-ton locomotive had to be capable of pulling a twenty-ton load at ten miles per hour, a five-tonner fifteen tons, and lighter locomotives proportionally lighter loads. Fifty pounds per square inch was the maximum steam pressure allowed.

The *Rocket* over which Robert Stephenson sweated during the summer of 1829 had inclined cylinders like the *Lancashire Witch.* But the wheels were not coupled. The pistons drove only the four-foot eight-inch front wheels. The smaller rear wheels did no work. It was the *Rocket's* boiler that was revolutionary: one of the first multitubular boilers ever fitted to a locomotive, and thus the precursor of the type used in most steam locomotives until the breed declined and disappeared.

Had you journeyed (by coach, of course) to the scene of the Rainhill Trials on that Tuesday in October, 1829, you might well have assumed that a horse race was about to take place. A grandstand bedizened with flags and bunting, an enthusiastic brass band, and a very busy tavern to help allay the thirst of the ten thousand spectators enlivened the course, which consisted of twin railroad tracks two miles long. The competitors with their racing colors were:

1. Messrs. Braithwaite and Ericsson of London, *The Novelty,* copper and blue, weight: 2 tons, 15 cwt. (hundredweight). John Ericsson, the Swedish-born inventor, would later build the *Monitor* for the Union navy during the American Civil War.
2. Mr. Hackworth of Darlington, *The Sans Pareil,* green, yellow, and black, weight: 4 tons 8 cwt., 2 qr. (quarterweight).
3. Mr. Robert Stephenson of Newcastle-upon-Tyne, *The Rocket,* yellow and black, white chimney, weight: 4 tons, 3 cwt.
4. Mr. Brandreth of Liverpool, *The Cycloped,* weight: 3 tons, operated by a horse.
5. Mr. Burstall of Edinburgh, *The Perseverance,* red wheels, weight: 2 tons, 17 cwt.

Brandreth's *Cycloped* was really not competitive, for it was propelled by a horse on a treadmill. Burstall's *Perseverance* was damaged when the wagon on which it was being brought to Rainhill upset en route. It therefore arrived late and was withdrawn. *Novelty,* with its vertical boiler and unusual configuration, was beautiful to behold in its brave blue paint and polished, copper-sheathed parts. And it was the crowd's favorite. In fact, it made a great show as it sped down the track at a phenomenal 28 mph. Sadly, however, a blow-back from its firebox burst the mechanical bellows with which it was fitted. Flames, smoke, and sparks erupted with a loud bang. *Novelty* was, at least temporarily, out of the running. *Sans Pareil's* boiler started leaking and Hackworth asked for more time in which to get it repaired. It was by now late in the day and raining hard. Stephenson was told his turn would come the next morning.

In the morning the *Rocket* was weighed, steam was raised to fifty pounds, and two wagonloads of stone and the tender were coupled up. Their weight was 12 tons, 15 cwt. George Stephenson himself took the throttle, with Robert alongside. After cautiously running up and back successfully according to the rules and making a time of four minutes and twelve seconds on its tenth run, Stephenson decided to show what *Rocket,* with success almost assured, could do on its last timed run. He opened the throttle wide. With a white plume of steam streaming from her tall stack and her piston rods a blur, *Rocket* roared past the grandstand, her wagons swaying behind. Time: three minutes, forty-four seconds— or twenty-nine miles per hour.

OHN STEVEN
1825

PETER COOPER'S "TOM THUMB" 1829-30 BALTIMORE & OHIO R. R.

Opposite: Peter Cooper's tiny locomotive *Tom Thumb* was the first to haul passengers on an American railroad. Below: Tom Thumb races a horse in 1830. The horse won. Bottom: Horatio Allen's *Best Friend of Charleston,* built in 1830 for the South Carolina Railroad.

Copyright—1883—by Hoopes and Townsend.

FIRST RAILWAY TRAIN IN PENNSYLVANIA,

Drawn by "Old Ironsides," the first Loco built in the U. S

FIRST TRIP, 23rd NOVEMBER, 1832.

Below: Model of *Old Ironsides.* Built in 1832 by Matthias Baldwin of Philadelphia for the Philadelphia, Germantown & Norristown, this remarkable locomotive ran a mile in the then astounding time of fifty-eight seconds.

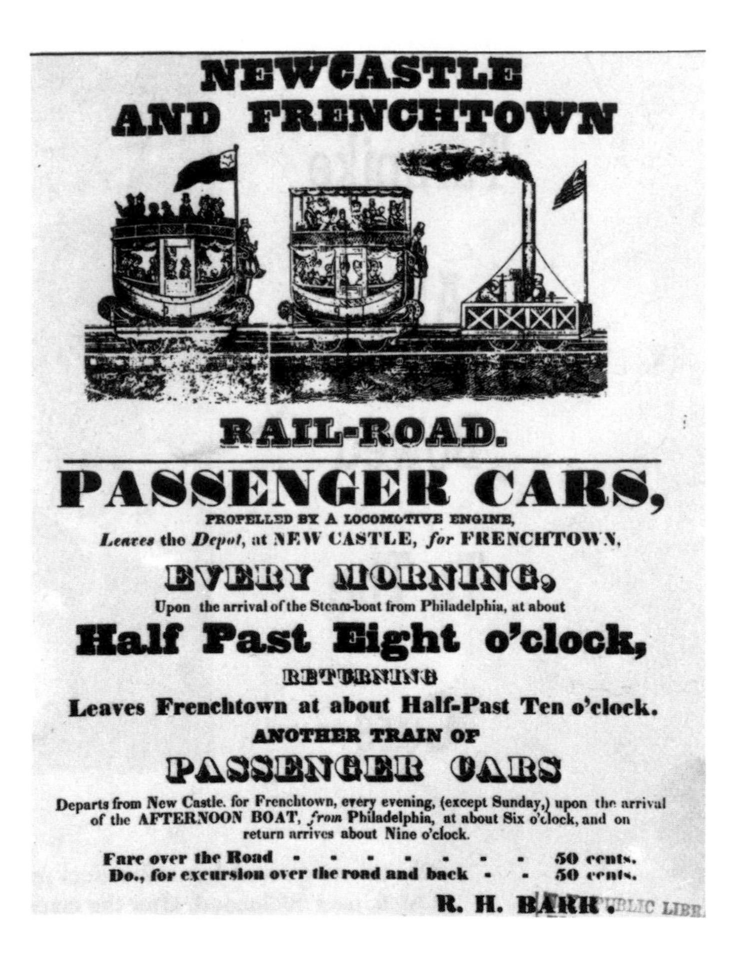

But the contest was not yet won. Many in the crowd still favored the pretty *Novelty*. George Stephenson wasn't worried. "Eh, mon, we needna fear yon thing," he said. "Her's got nae goots."

Novelty's sponsors asked for more time to get ready. A few days later she had her chance. She made only one run before the pipe from her feed pump to her boiler burst, "water flying in all directions." Nor after repairs did she succeed in approaching the *Rocket's* performance. Hackworth's *Sans Pareil* did even worse.

As might be expected, four *Rocket*-type engines were ordered for the Liverpool & Manchester. Strangely, enthusiasts for the *Novelty*-type still existed on the board of directors and two such were ordered in spite of their higher cost. A *Novelty* cost £1,000, a *Rocket* only £650. One thing was certain. Even the most mossbacked of railway directors no longer thought that trains ought to be horse-drawn.

The Liverpool & Manchester railway—all thirty miles of it—opened for operation in September, 1830. Public opinion now swung in favor of this spectacular new transport. Wildly enthusiastic throngs admired the tunnels, the bridges, the stations and were as impressed with the locomotives and rolling stock as were those mobs which used to haunt Cape Kennedy to admire the aerospace rockets.

The opening ceremonies were disastrous. That iron-headed prime minister, the Duke of Wellington, had been invited. Wellington, the hero of Waterloo, was a right-wing Tory and violently against the reform of Parliament, one of the big issues of the time. At his appearance, the crowd rioted. Further, William Huskisson, a "liberal" Tory whom Wellington deigned to greet while the bigwigs alighted from their special car, was run over by a train operating on a parallel track. This tragic accident resulted in a spectacular run by George Stephenson on the locomotive *Northumbrian* to get Huskisson to a doctor. The then incredible average speed of thirty-six miles an hour was maintained for fifteen miles and seems to have caused more of a stir than poor Huskisson's death that night.

The next day the first scheduled passenger train on the L&M left Liverpool bound for Manchester. Tickets were bought by one hundred and forty passengers. During the next century the thirty miles of fragile track in northern England increased to hundreds of thousands of railway miles on every continent.

British engineers were not the only ones to build and improve locomotives, track, and rights of way. Their counterparts in other countries fussed eagerly with pistons, cylinders, and boilers, and with track and signals, too.

In France the great Marc Seguin tested an advanced locomotive for the St. Étienne-Lyons Railroad even as the *Rocket* was winning the Rainhill Trials. This unusual engine, the first built in France, not only had a multitubular boiler, but also huge, axle-operated fans mounted on its tender. These blew air under the fire by means of flexible leather tubes, thus increasing its intensity.

The United States had its own railroad geniuses. One was John Stevens, the steamboat man, who as early as 1812 had published a pamphlet urging a steam railroad instead of the Erie Canal then under consideration. "To the rapidity of the motion on these railways no limit can be set," he wrote. "I can see nothing to hinder a steam carriage from moving on these ways at one hundred miles an hour." For an era whose fastest transportation was a running horse, this was a remarkable prediction, indeed.

Stevens had designed a steam locomotive as far back as 1795, but it was a huge, low-pressure device much too heavy for any track. In 1817 the state of New Jersey granted him a charter "to build a railroad from the river Delaware, near Trenton, to the river Raritan near New Brunswick." No railroad was built, but Stevens was preparing the way. In 1823 he projected a railroad from Philadelphia to Pittsburgh—the Pennsylvania Railroad Company. In 1826 he built the first locomotive actually to run on rails in America. It operated on a

Opposite: Poster of the Newcastle & Frenchtown Railroad, showing cars that still look like horse-drawn coaches. The locomotive depicted has a vertical boiler, much like that of Peter Cooper's *Tom Thumb.*

circular track near the present Stevens Institute in Hoboken. Three years later Horatio Allen ran the British-built *Stourbridge Lion* on the tracks of the Delaware & Hudson Canal Co., which was used to haul coal cars from Pennsylvania mines to the canal. But the seven-and-a-half-ton *Lion* was too heavy for the crude tracks and trestles, and it was soon withdrawn.

These beginnings, feeble as they were, encouraged a rush toward steam railroading. While the too-heavy *Stourbridge Lion* was pounding its tracks to shards, Peter Cooper, the American mechanical genius, built a locomotive, the *Tom Thumb,* at the other extreme—too light and tiny. His engine weighed but one ton, had one cylinder, and developed not much more than one horsepower. It had a tubular boiler, and since American foundries were as yet unable to fabricate such pipes, Cooper used musket barrels. Still, in 1830, *Tom Thumb* was the first engine to pull a trainload of American passengers on an American railroad—the Baltimore & Ohio. One of Cooper's trips has long been a part of railroad folklore because he raced his engine against a horse on a parallel track. The horse won.

Another famous early American engine was *The Best Friend of Charleston,* designed by Horatio Allen for the South Carolina Canal & Railway Company after Allen had quit the Delaware & Hudson. It was built at the West Point foundry, whose usual business was the casting of cannon. It hauled forty passengers in four cars at twenty-one miles an hour. Later it won the distinction of being the first American engine to blow up. Its fireman, annoyed by the loss of pressure when the safety valve popped open, sat on said valve to his lasting regret. On later trips a car loaded with cotton bales was coupled on behind the tender as a safety measure.

Horatio Allen then designed an engine called the *South Carolina,* which suited American conditions better than machinery imported from England. In England, the Stephensons had established a tradition of building rights of way with long sweeping curves. American lines twisted more and British-style engines with their rigid axles tended to jump the tracks. Allen pivoted his front truck, an invention which for years was peculiar to all American engines.

The first really successful American locomotive was built by Matthias Baldwin of Philadelphia. Baldwin had been a watch repairer and then a builder of machinery for printing calico, and had built a small steam engine to power his shop. When asked to build a locomotive for the Philadelphia, Germantown & Norristown road, he took a long look at the *John Bull,* an imported Stephenson-built engine for Stevens' Camden & Amboy R.R., and then constructed *Old Ironsides,* which in 1832 put up the astounding performance of a mile in fifty-eight seconds. Other American makers soon jumped into the business of building locomotives suited to American needs—Rogers, Hinckley, Norris, and the rest. And soon American-built engines were being loaded aboard ships to run on rails all over the world.

America also showed the world how to lay those rails quickly and cheaply. And to the discomfort of the passengers. John Stevens' son Robert wanted neither the iron-plated wooden or stone—yes, *stone*—rails Americans had been experimenting with. Nor did he wish to use the expensive system of mounting the rails on chairs set on stone blocks which the Stephensons used. He whittled a type of rail which could be quickly spiked to wooden cross ties. This broad-based "T" rail became the world's standard rail, although the British and some others still use chairs instead of the hook-headed American spike to fasten it to the ties (which the English call "sleepers").

When Richard Trevithick ran his *Catch Me Who Can* around its little circle of track, he had no idea that the railway would stretch the narrow world of the eighteenth century across prairies and deserts and continents, and spread it into areas never before accessible to man and his goods. Nor did he dream, in his penury, that later generations of railway promoters, especially the grand American buccaneers of the nineteenth century—the Goulds, the Fisks and their ilk—would find in railway speculation the riches of Golconda. And the Stephensons, who had perhaps a longer vision, would stir in their graves if they knew that the railways which had waxed so great for so many generations had at last almost succumbed to powered road carriages not too unlike those of Goldsworthy Gurney.

3

3 Taming the Lightning

Electricity Comes of Age

Before any of our modern means of electrical communication—the telegraph, the telephone, radio, and television—could be made to work, electricity had first to be understood and then controlled and applied to technological problems.

The ancients had amused themselves by rubbing pieces of amber, which then attracted bits of feather and other susceptible objects. But it wasn't until 1600 that William Gilbert of Colchester, England, physician to Queen Elizabeth, showed that other substances besides amber could be made to do the same trick, and that the effect was evidence of electricity—static electricity, as we now know—at work. It was Gilbert who gave us the word *electrica,* from the Greek word for amber, *elektron.*

In 1672 Otto von Guericke devised a mechanical means of creating electrical effects. He mounted a large sphere of sulfur on a rotatable shaft. When he rubbed the spinning sulfur ball with his hand, the ball became capable of attracting and repelling bits of paper. In 1709 Francis Hauksbee improved on von Guericke's device. His friction electric machine used a rapidly revolving glass globe against which he held his hand. An iron chain touched the spinning sphere and conducted the electrical charge up to a horizontal gun barrel hung from the ceiling. At the other end of the barrel was another chain. When a brave person put his or her finger near the end of the chain, a bright crackling spark jumped from chain to finger.

This display of drawing-room lightning fascinated rich eighteenth-century "philosophical" experimenters, and soon it became the in thing to have an elaborate electrical machine with which to amaze one's visitors. The machines were rapidly improved during the century, and revolving balls gave way to discs which spun between rubbers coated with tin amalgam and with comb-like conductors for drawing off the charge.

In 1745 a means of storing the charge in a condenser was discovered by Petrus van Musschenbroek in Leyden, Holland. Musschenbroek had been trying to electrify water with a Hauksbee-type machine. A student held the jar of water into which the chain from the machine hung down. After the charging had been going on for a while he happened to touch the chain. The electric shock almost killed him.

Soon afterward, Musschenbroek discovered that a dry bottle with the lower part of the inside and outside surfaces coated with tin foil worked better than a hand-held bottle of water. This became known as the Leyden jar.

The Leyden jar created a sensation all over Europe. Electric shocks became a fad. At the French court a line of one hundred and eighty guardsmen held hands and leaped into the air in formation when they received the charge from a jar. Not to be outdone, seven hundred French monks tried the same experiment and duly jumped heavenward.

When Benjamin Franklin saw a similar demonstration, he couldn't wait to get his hands on an electric machine and a Leyden jar. His experiments proved to be of great importance in advancing the understanding of electricity.

Before Franklin made his investigations it was thought that electricity was a "fluid" that came in two forms. He claimed that there was but one kind. He also showed that the electricity of lightning was the same as the electricity generated by an electric machine. It was in 1752 that Franklin conducted his much-celebrated experiment, flying a kite during a thunderstorm. He charged Leyden jars from a key attached to the kite string, as the lightning crashed overhead. (He was lucky to escape with his life; another experimenter following his lead in 1753, the Russian physicist Georg Wilhelm Richmann, was killed in St. Petersburg.) It was during these experiments that Franklin invented the lightning rod.

The discharge from a Leyden jar is sudden and complete. Early telegraphic experimenters tried using such a sudden current for sending messages over wires, although what was needed was a continuous current. The work of the Italian anatomist Luigi Galvani led in that direction.

In 1780 Galvani, studying the nervous systems of frogs, found that if he pushed a brass wire into a dead frog's

Preceding pages:
Primitive painting of
Franklin's lightning experiment.
1. Electric machines amused
eighteenth-century aristocrats.
2. Glass cylinder
electric machine, complete with
Leyden jars and other gadgetry.
3. Some lightning rods
went to ridiculous lengths.
4. Benjamin Franklin.
5. Russian scientist Richmann
tries Franklin's experiment and
electrocutes himself.

1

2

3

4

5

1. Luigi Galvani.
2. Galvani performs his frog experiment.
3. Alessandro Volta.
4. The first electric battery, the Voltaic pile.
5. American electrical genius Joseph Henry.
6. Henry's experimental electric motor of 1831.
7. Michael Faraday.
8. Pacinotti's dynamo of 1860.

1

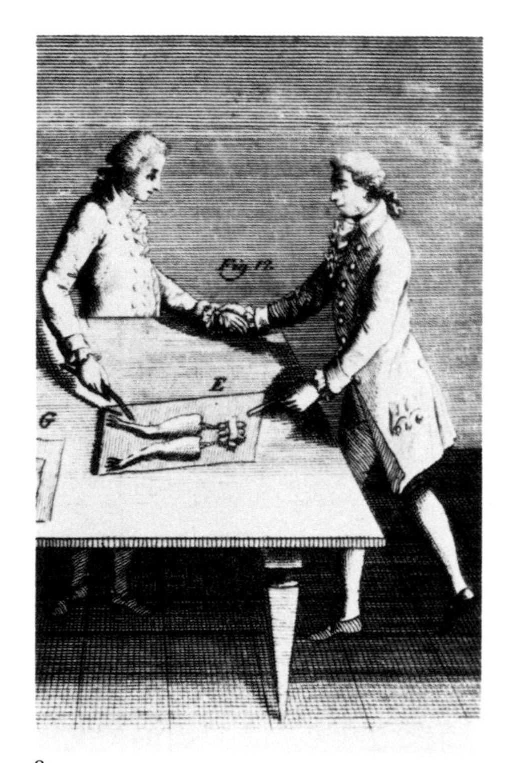
2

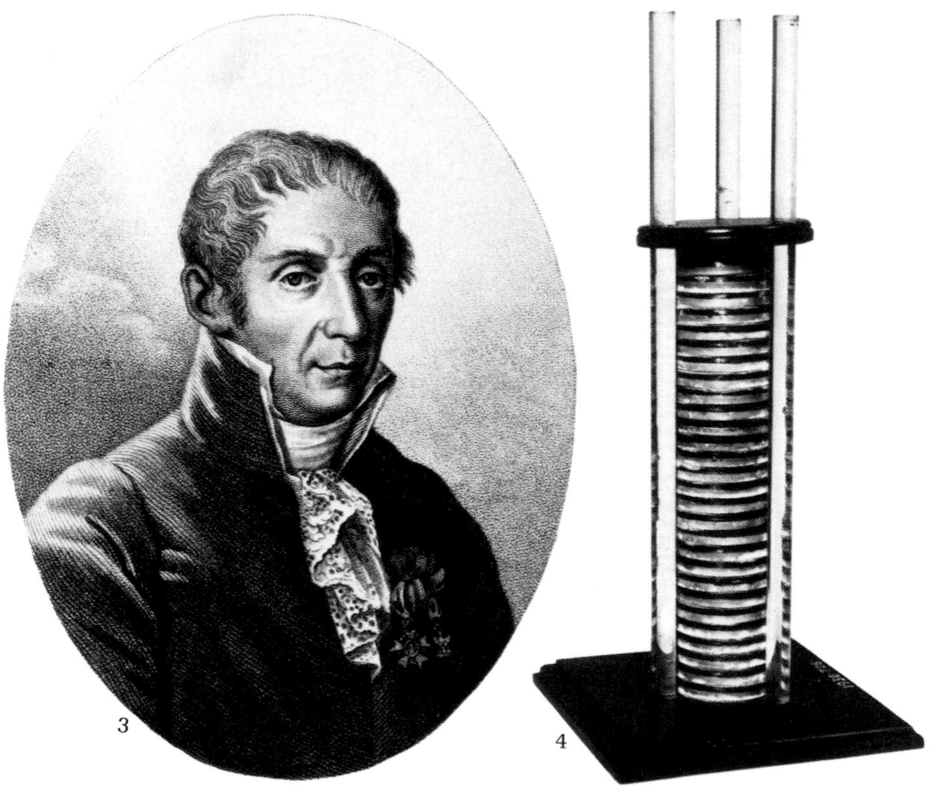
3

4

5

6

7

8

spinal marrow, and then allowed the wire and the frog's feet to touch an iron plate at the same instant, the frog's legs jumped violently. They were, as we now say, galvanized into action. Galvani thought that when the wire and the plate came in contact there was a flow of electricity from the frog's nerves to its muscles, which caused the muscles to contract.

Alessandro Volta of Pavia, Italy, didn't accept Galvani's theory. He rightly assumed that there was a flow of electric current caused by the connection of dissimilar metals through a moist saline conductor—the dead frog. In 1800 he put together a cylindrical pile of copper and zinc discs separated by moistened paper discs (later the paper discs were moistened with dilute salt or acid solutions). When he touched the top disc with one hand and the bottom disc with the other, he felt a distinct shock. But unlike the shock from a Leyden jar, this sensation continued until he removed his hands. He had built the world's first electric battery, the Voltaic "pile." Most importantly, because it supplied a steady current, rather than an evanescent flash or shock, Volta's battery enabled scientists to study electricity and expand their theoretical knowledge of it. The phenomenon of electrolysis was soon discovered, and through it Sir Humphrey Davy was enabled to isolate the fortieth and forty-first elements, potassium and sodium. Electrolysis also found practical application in the electroplating of metals.

Over the following decades, improved versions of the electric battery were developed, but none was rechargeable and the electricity they produced was extremely expensive. The accumulator or storage battery appeared in 1859; further improved, it became the lead-acid battery still used in the world's automobiles.

It was the dynamo, generating electricity by moving wire coils in a magnetic field, that eventually became a constant, reliable, and economical source of energy. By the 1870s power could be transmitted over long distances and used to illuminate single elements such as lighthouses and searchlights and, within another decade, entire cities and nations.

The dynamo was one result of the discoveries in magnetism and induction by Joseph Henry in the United States and Michael Faraday in England. Henry, in 1829, and Faraday, in 1831, found that electricity could be produced by magnetism—the opposite effect of Hans Christian Oersted's discovery, in 1820, that magnetism could be obtained from electricity.

The discovery of electromagnetic induction inspired a long line of experimenters to try to build current-generating machines. The first practical dynamo, built in 1860 by the Italian physicist Antonio Pacinotti, created little excitement. But the response was different ten years later, when a Frenchman named Zénobe Théophile Gramme came up with a machine similar to Pacinotti's. Because Gramme was backed by businessmen who knew how to promote his machine, his dynamo became a commercial success. By 1890 generators were almost as efficient as those of our time.

The early telegraphs were powered by wet-cell, nonrechargeable batteries. The telephone at first also used batteries. Today's radio and television stations receive their current from power lines fed by huge dynamos, some of them

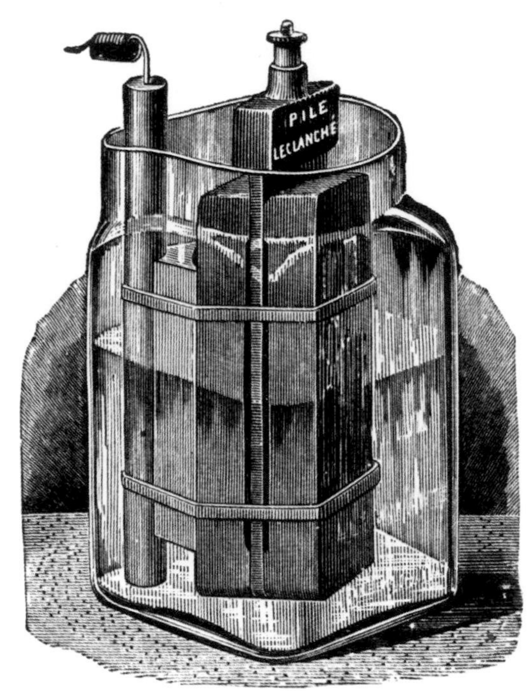

powered by atomic fission.

The increasing sophistication of power sources parallels the continued movement toward greater sophistication in electrical communication methods.

The telegraph was an incomparably faster method of delivering intelligence than anything that preceded it. But the telephone was even quicker and more direct. The words spoken into the telephone transmitter went down the wire whole; they didn't have to be broken down into dots and dashes and then reconstituted at the other end of the wire. The telephone was also more personal; the people on the line could hear the nuances of speech rather than having to read a curt message on a telegraph form.

Both the telegraph and telephone had a tremendous effect on people's lives. But their effect was minimal compared with that of radio and television. For these two, especially television, not only disseminate information, they amuse, chagrin, educate, and inspire the entire population of the planet. Heroes are created, political leaders are elected, the fate of nations is decided, all through sounds in a loudspeaker or dots on the face of a tube.

What is the next step? Will it be that devilish device that's been murmured about recently whereby sounds and pictures (in glowing color) will appear inside our heads as we hold a cigarette-like electrode between our lips?

Let us pray it is not.

Electric Dots and Dashes

Samuel Finley Breese Morse was not the inventor of the electric telegraph. God wrought it for other men before him. But Morse's system turned out to be the most practical.

The word "telegraph" was first applied to the semaphore signal systems of mid-seventeenth-century Europe. Masts or towers, set on hills within telescope range of each other, relayed messages by means of wooden arms which could be swung into various positions to represent letters. The British used telegraphs to report the arrival of ships in the Channel, and an elaborate French system could transmit messages across the country faster than they could be carried by post riders.

In the United States, Cape Cod still has "telegraph hills" that were used in 1800 to receive and pass along signals from Martha's Vineyard concerning incoming ships. Telegraph Hill in San Francisco served a similar purpose for ships coming into the Bay. Semaphores, of course, were simply refinements of the age-old system of hilltop signal fires by which the Romans sped messages over the Alps to Gaul and Britain, or a visual counterpart of the drumbeat message relays of tribal Africa.

It took about a hundred years for telegraphy to become useful—from the time in 1747, when a curious English cleric, a Dr. Watson, bishop of Landaff, stretched a wire across the Thames and by means of a Leyden jar gave an electric shock to himself and to an accomplice dipping an iron rod into the river on the other bank. Later that year the doctor transmitted an electric impulse through 16,000 feet of wire suspended between poles. Benjamin Franklin performed a similar feat across the Schuylkill River in Philadelphia. Neither Franklin nor Watson thought of sending a suitably interrupted current down a wire in order to convey a message. But almost every other early experimenter with electricity seems to have taken a crack at it.

At first, the only sources of electricity for experimentation were the so-called "electric machines," which were friction generators, or Leyden jars. (The electric battery did not become available until Alessandro Volta invented it in 1799.) The first telegraph transmission system—in Geneva, a quarter-century earlier—used the Leyden jar as a power source. The inventor strung up twenty-four insulated wires, one for each letter of the alphabet (not all alphabets have twenty-six letters). From the receiving end of each wire was suspended a pair of pith balls. When a charge was sent down a particular wire, the appropriate pith balls were repelled by each other and thus indicated which letter was being transmitted.

Opposite, left: Zénobe Théophile Gramme's dynamo of 1870. Although Gramme's machine came ten years after Pacinotti's, it was more vigorously promoted and achieved greater success. Right: Batteries in open jars filled with acid were the chief source of electrical power until the dynamo came into use.

One of the first men to use a battery—a Voltaic pile—for telegraphy was an S. T. Sommering, who communicated his invention to the Academy of Sciences in Munich in 1808. His system depended on the recent discovery that water could be decomposed into hydrogen and oxygen by electricity. He used thirty-five wires and thirty-five water-filled glass tubes each marked with a letter or numeral. The tubes receiving the electric current bubbled and thereby indicated which letter was being sent. Herr Sommering also proposed a rather nerve-rackingly loud system for alerting the receiving station that a message was about to be sent: exploding an accumulation of hydrogen with a spark.

Other people fiddled with systems using the principle of electrical decomposition of chemicals. One had some success with a thirty-six-wire rig. Another operated a single eight-mile wire on a Long Island race course. He used litmus paper on which red marks appeared when current flowed.

H. C. Oersted, a Dane, then discovered that a magnetic needle could be deflected by an electric current passing through a nearby wire; that it could be deflected in opposite directions according to the direction of the current. Then it was found that the needle's deflection could be increased by coiling a wire around the needle.

André Marie Ampère, whose name is still a term of electrical measurement, saw the magnetic needle as a means of telegraphy and in 1820 proposed that a lettered keyboard at one end of the line be used to control a separate needle for each letter at the receiving end. It failed, mostly because of the sixty wires required. But the magnetic-needle principle became the basis for a dizzying number of telegraphic systems, some of which were used successfully well into the 1840s.

In 1833, Counsellor Karl Friedrich Gauss and Professor Wilhelm Eduard Weber built such a system over the houses and steeples of Göttingen. They had constructed an iron-free magnetic observatory where they made observations and from which they sent messages over a three-mile line built for their experiments in electrical physics.

Gauss was one of the great masters of modern mathematics and propounded important theories on earth magnetism and optics. Weber was the physicist who first showed that it was possible to define electrical quantities. He, too, did some of the first important research on magnetism. Telegraphy, radio, and television owe much to the basic work of these two men.

Professor Carl August von Steinheil of Munich much improved their system of telegraphy. He succeeded in reducing it to a single wire—the return circuit was through the ground. He also used a system of dots and dashes impressed on moving tape, similar to those which were later used by Morse. Steinheil's telegraph was working by 1837 and was adopted by the Bavarian government a year later.

The most practical and most used of the needle telegraphs was that patented in England by Charles Wheatstone and William Fothergill Cooke in 1837. Wheatstone in 1835 first measured the speed of electricity (he got it wrong) and, among many other things, invented the stereoscope.

Wheatstone and Cooke's needle telegraph depended on the swinging of five needles actuated by five lines of wire. The needles pointed at letters on a diamond-shaped board. Two needles pointed at each letter. There were only twenty letters; C, J, Q, U, Z were omitted. Later it was found that five needles were unnecessary since the operators soon learned to read the message from the angles of the needles and disregarded the letters. Only two needles were used in later instruments, and still later only one.

The Wheatstone-Cooke telegraph was installed in 1842 on the London & Birmingham and the Great Western railroads to control traffic. Several years later the English public had a vivid demonstration of telegraphy's effectiveness when a message from the railway station at Slough to Paddington led to the capture of an escaped murderer.

The further evolution of the telegraph, however, followed a direction made possible by the electromagnet. Invented in England in 1825 by William Sturgeon, it was a piece of soft iron bent into the shape of a horseshoe and wound with wire. If an electric current was passed through the wire the horseshoe would attract and hold another piece

1

4

5

1. Claude Chappe's pre-electric hand-operated semaphore telegraph was used in France. 2. Georges Lesage's telegraph of 1774 used an "electric machine" for current. 3. Model of Sommering's telegraph of 1808, powered by a Voltaic pile. 4. Hans Christian Oersted. 5. Karl Friedrich Gauss and Wilhelm Eduard Weber.

2

3

1. William F. Cooke, who suggested the idea of the telegraph to Wheatstone. 2. Charles Wheatstone. 3. Four-needle Wheatstone-Cooke telegraph. 4. Exaggerated French engraving of a two-needle telegraph. 5. Wheatstone-Cooke two-needle telegraph used in the British House of Commons. 6. Advertisement for the Wheatstone-Cooke telegraph.

1

2

of iron. The great American physicist, Joseph Henry, increased the electromagnet's power by winding it with nine sixty-foot wire coils. In 1831, in Albany, New York, he built and worked a mile-long telegraph which used a ten-inch steel bar mounted on a pivot. One end of the bar was between the poles of an electromagnet and was moved by the current moving through the magnet. The other end rang bells or made dots on paper to indicate letters. Five years later, at Princeton, New Jersey, Henry ran "the first actual line of telegraph using the earth as a conductor [instead of a second wire] . . . signals were sent . . . from my house to my laboratory."

But Henry didn't patent his telegraph, although he did publish papers describing it. Later he explained, with regret, that "I declined [to apply for a patent] on the ground that I did not then consider it compatible with the dignity of science. . . . I was too fastidious."

Early telegraph systems were beautifully constructed by noted physicists who knew all there was then to know about electricity and magnetism. Yet it was Samuel F. B. Morse, a fair painter, a so-so daguerreotypist, an ignoramus about electricity, and a nearly incompetent mechanic, who succeeded in giving the world a practical telegraph system—and got rich doing it.

Morse was born in 1791 in Charlestown, Massachusetts, the son of a rigidly Calvinist pastor who was also a stiff Federalist. At seven, Finley Morse was shipped off to Phillips Academy at Andover and at fourteen he entered Yale. Here he was no great shakes as a student, but made his first feeble attempts at drawing and painting, and was marginally initiated into the mysteries of the "electric fluid." Professors Jeremiah Day and Benjamin Silliman were the scientific doyens of Yale in those early years of the nineteenth century and from them Morse got his smattering of electrical and chemical knowledge. Day demonstrated the visibility of "electric fluid" by passing it through a chain in a darkened room. The "fluid," in the form of sparks, could be seen between the links. He also showed that the sparks could perforate paper. And he had his class join hands in a circle to receive a simultaneous shock. The high-frequency current for these games came most likely from a Leyden jar. But later in Silliman's class Morse saw a new source of electricity, Alessandro Volta's "pile," the first battery. Not only was Morse allowed to examine this wonder of science, he was even permitted to take it apart so that he might copy it to delight his awed friends. For manipulating "electric fluid" was an awesome thing back then—before Napoleon had even thought about capturing Moscow—perhaps more awesome than a contemporary college sophomore doing a little atomic fission in his room.

Mostly Morse pleased his cronies in New Haven by drawing caricatures and portraits. And soon, with a Yankee's keen appreciation of a buck, he was charging for portraits: $1 for a profile, $5 for a miniature on ivory. It was a lazy life, and when Morse left Yale in 1810 he had not prepared himself for a profession. Since the last recourse of incompetents often is art, Morse turned to painting. Two of the country's best-known painters, Gilbert Stuart and Washington Allston, who worked abroad, thought enough of the work he showed them to convince his parents that Morse might at least be made into an artist. So in 1811, when Allston took ship for London, Finley Morse, aged twenty, went along to become his student.

A year later England and the United States were at war. But wars were more gentlemanly in those days and no one interned him behind barbed wire (the stuff hadn't been invented yet). Morse had a fine time in London, fell in with the art establishment, watched a balloon ascension, got to know the great Benjamin West, and even exhibited at the Royal Academy.

By 1815 he was back in Boston, confidently awaiting the throngs who would importune him for paintings. No one importuned him. He headed for the hinterland to peddle portraits. Surprisingly, he was fairly successful even though in London he had been painting large and rather dull compositions devoted to classical and historical subjects, and had looked down upon mere portraitists. But being an itinerant artist palled.

His next money-making project was a pump which he and his brother Sidney devised. An improvement on the

3

4

5

Under the Special Patronage of Her Majesty

And H. R. H. Prince Albert

GALVANIC AND MAGNETO

ELECTRIC TELEGRAPH,

GT. WESTERN RAILWAY.

The Public are respectfully informed that this interesting & most extraordinary Apparatus, by which upwards of 50 SIGNALS can be transmitted to a Distance of 280,000 MILES in ONE MINUTE,

May be seen in operation, daily, (Sundays excepted,) from 9 till 8, at the

Telegraph Office, Paddington,

AND TELEGRAPH COTTAGE, SLOUGH.

ADMISSION **1*s*.**

"*This Exhibition is well worthy a visit from all who love to see the wonders of science.*"—MORNING POST.

Despatches instantaneously sent to and fro with the most confiding secrecy. Post Horses and Conveyances of every description may be ordered by the ELECTRIC TELEGRAPH, to be in readiness on the arrival of a Train, at either Paddington or Slough Station.

The Terms for sending a Despatch, ordering Post Horses, &c., only One Shilling.

N.B. Messengers in constant attendance, so that communications received by Telegraph, would be forwarded, if required, to any part of London, Windsor, Eton, &c.

THOMAS HOME, *Licensee.*

G. NURTON, Printer, 48, Church Street, Portman Market.

6

common force pump which was joked about as "Morse's Patent Metallic Double-Headed Ocean-Drinker and Deluge-Spouter Valve Pump-Boxes," it was meant to be used on hand fire engines, boats, and blacksmiths' bellows. The pump was, finally, a financial flop.

In 1818 he married Lucretia Walker of Concord and took to the road again with his canvases and brushes, this time to Charleston, South Carolina. For a while he prospered, then business fell off. At home in Massachusetts disaster piled on disaster. His father was forced to resign as pastor and the family moved to New Haven; Lucretia and Finley's baby died. Morse came north to New Haven. There he painted a few portraits—notably of Eli Whitney and Noah Webster. And he invented a marble-carving machine to reproduce sculptures, coincidentally an idea James Watt also was working on.

Frantically, Morse turned from one project to another. He went to Washington to paint a giant picture of Congress. He went to New York, then as now the most important art center in the country, to exhibit it. He charged admission. No one came. Again he hit the road as an itinerant limner of back-country physiognomies.

In 1825 his wife died, in 1826 his father, in 1828 his mother. During those years he had hustled and bustled, enjoying some success in New York; he had formed and become head of a National Academy and had painted a well-received portrait of Lafayette. He spent the next three years in Europe and England, part of the time furiously painting away at a monstrous big canvas of the Louvre showing dozens of famous paintings crammed on its walls.

Returning home in 1832 in the sailing packet *Sully*, he became involved in a conversation in the dining saloon one night about Ampère's experiments with electricity. One of the men at the table wondered if the flow of the "electric fluid" would be diminished if it traversed a very long wire. A windy young doctor, Charles T. Jackson, assured the company that electricity could go on for miles and miles. Morse became excited. "If this be so," he said, "and the presence of electricity can be made visible in any desired part of the circuit, I see no reason why intelligence might not be instantaneously transmitted by electricity to any distance."

Morse thought he had a brand new idea. He didn't know that Lesage in Geneva had tried it years before he was born. He stayed up late that night making drawings. He worked out a rough code and he drove his fellow passengers crazy, especially Doctor Jackson, with his endless talk about his great new electric telegraph.

Upon arrival in New York he immediately immersed himself in the construction of his brainchild. Within a week the telegraph was put aside. Mundane worries, like making a living, intruded. He wanted to finish his painting of the Louvre. He became busy teaching art at the University of the State of New York. He also wasted a lot of time on nutty politics.

Morse was a zealot of the far-right lunatic fringe. He hated "papists" and immigrants, especially the Irish, who were both. In 1836 he ran for mayor of New York as the candidate of the Native American Association. He lost. He got only 1,496 votes. He went back to the telegraph.

Morse's transmitting device was not the familiar telegrapher's key of later years. At first it was based on a typesetter's stick. Instead of type, each letter was cast in lead in the form of bumps, each corresponding to a dot or dash. The bumps acted like cams to move a pivoted stick, one end of which acted like a switch. His receiver was made out of an old painter's canvas-stretcher nailed vertically to the side of a kitchen table. His magnets were mounted on a wooden bar fastened midway down the stretcher. These attracted and repelled a sort of pendulum to whose lower end was attached a pencil. An old wooden clockwork pulled a paper tape under the pencil which drew an angular wavy line which could be read as dots and dashes.

The power for the electromagnets was supplied by crude wet-cell batteries which he most likely made himself from strips of zinc and copper immersed in jars of acid. (It was, of course, impossible in New York in 1836 to go to a shop to buy batteries, magnets, or insulated wire. In London or Paris perhaps, but not New York.) Morse was not a very knowledgeable constructor of such components. When he first put together his device and energized the system it just

Opposite: American physicist Joseph Henry built this electromagnetic telegraph apparatus in Albany, New York, in 1831. With it he worked a mile-long telegraph. In the instrument shown, the pivoted iron bar rang an electric bell.

lay there—not a single buzz of life. He diddled with this wire, poked that one. Nothing happened. He begged one of the university physicists, Professor Leonard Gale, to take a look. Gale, who was familiar with Henry's work, was appalled. Morse had wound bare, uninsulated wire on his magnets. Not until he was shown how magnets and batteries ought to be hooked up did he get any results.

Morse, abysmally ignorant of what had already been done in telegraphy, and astounded that the idea was not original with him, realized that he needed Professor Gale to perfect his system and almost immediately made him a partner. Gale filled him in, calling his attention especially to the work of Professor Henry, who was a friend of Gale's.

After making changes suggested by Gale, Morse was able to send messages, first through a hundred feet of wire, then through a thousand, and in 1836 through reels on which were wound ten miles of wire.

Morse's next partner was Alfred Vail. Vail was of the same peculiar political persuasion as Morse; he had been a student at the university where Morse taught and had watched Morse's experiments there. Vail, a good mechanic, was the son of the owner of the prosperous ironworks in Morristown, New Jersey, which later became the Baldwin Locomotive Works. Fascinated by the telegraph and convinced of its great potential, he offered to help Morse with money (his father's) and mechanical assistance. He also undertook to build a patent model of the telegraph. In return he was to get a one-fourth interest in the invention. Enlisting young Vail was one of the two smartest things Morse ever did. The other, of course, had been enlisting Gale. In years to come, Vail made Morse's clumsy code workable, invented the telegraph key to replace Morse's ridiculous composing stick, and also devised a printing telegraph. Under the terms of Vail's contract with Morse all of Vail's improvements were patented in Morse's name.

Morse now entered a caveat for a patent, a specification of what he intended to patent when it was completed. Short of money, as usual, he got Vail to ante up the fee—$30. On October 6, 1837, the U.S. Patent Office acknowledged that the caveat had been recorded.

Earlier, the Congress had asked Levi Woodbury, President Van Buren's Secretary of the Treasury, to report on the feasibility of building a telegraph system. The congressmen were thinking of semaphores; they hadn't heard about electric telegraphs. Five proposals came in, four of them semaphore systems and one—Morse's—electric. In February, 1838, at the invitation of the House Committee on Commerce, Morse and Vail took the telegraph to Washington. It created a sensation. It created an even greater sensation in the agile brain of the House Committee's chairman, one F. O. J. Smith, a congressman from Maine and a very slippery customer, indeed. Morse would have been far better off if he had not met this swindler. Instead, Smith became Morse's third partner in the telegraph. His task: to con Congress into appropriating money for it. His agreement with Morse was signed in March. Almost immediately, "Fog" Smith sponsored a bill to appropriate $30,000 to build fifty miles of telegraph. He did not tell Congress that he had a finger in the pie, but he did claim that Morse had a patent. Which he didn't.

Morse and his crooked congressional confrere then set sail for Europe to patent the telegraph in as many foreign countries as possible. They did not have much luck. In England, the Attorney General, Sir John Campbell, refused Morse a patent on the grounds that the London *Mechanics Magazine* had already published details of it. (In England, publication precluded a patent.) Further, Wheatstone's apparatus was already in use, and Wheatstone most charmingly showed him how it worked. In Paris, Morse was dismayed to find that Wheatstone already had a French patent. Steinheil's telegraph was a working reality in Bavaria. The Russian Czar, Nicholas I, turned him down because the rapid dissemination of intelligence over wires might encourage subversion.

At least one nice thing happened to Morse in Paris. Daguerre showed him his daguerreotype and, as noted, the first news of daguerreotypy, which was to become a craze in America, came through Morse.

He returned to New York broke. The deep economic depression that followed the crash of 1837 still

Opposite: Samuel Finley Breese Morse, photographed in London.
Left: French engraving which purports to show Morse aboard the *Sully* in 1832, where he first got the idea for his telegraph.

continued. There was no money for congressional excursions into telegraphy. Morse grasped the daguerreotype as a drowning man grasps at straws. As soon as Daguerre's manual describing the process came to New York Morse became a disciple and then, to make some money, a teacher. Among others he taught the daguerrian art (at $25 to $50 per course) to Edward Anthony, who later founded the photographic business which became Ansco and is now GAF. And he started the great Mathew Brady on his way.

Despite what he could eke from lessons in photography and painting Morse remained terribly poor. One of his painting students, who later told this story, was behind in his payments.

"Well, Strother," Morse asked, "how are we off for money?"

"Why, Professor, I'm sorry to say I've been disappointed, but I expect a remittance next week."

"Next week?" asked Morse. "I shall be dead by then."

"Dead, sir?"

"Yes, dead by starvation."

"Would ten dollars be of any service?"

"Ten dollars would save my life. That's all it would do."

Strother took Morse to dinner and gave him ten dollars, as well. Morse said: "This is my first meal in twenty-four hours. Strother, don't be an artist. It means beggary."

Nor did it make Morse feel any better when Wheatstone and Cooke were granted a patent for their telegraph on June 12, 1840. Morse's patent didn't come through until June 20.

More than four years had gone by since Morse and "Fog" Smith had tried to get Congress to appropriate the money for an experimental telegraph system. Late in 1842 he went to Washington to try again. He was allowed to set up his equipment and string his wires in the rooms of the House Committee on Commerce and the Senate Committee on Naval Affairs, and send messages between them. The yahoos of Congress were again amazed and delighted. Then, day after day Morse sat in the House gallery waiting for the appropriation to be voted on. In February of 1843, the House of Representatives considered the bill to appropriate the $30,000 for the telegraph. Cave Johnson of Tennessee rose to suggest that since Congress was going to encourage the science of electromagnetism it ought also encourage Mesmerism. He thought half the appropriation ought to go to a Mr. Fisk, then experimenting with "animal magnetism." He offered such an amendment. George S. Houston of Alabama spoke in favor of also supporting the Millerites, a religious group which predicted the Second Coming of Christ in 1844. Although the legislators insisted that they were merely "funnin'," the chair ruled the motions in order—that "it would require a scientific analysis to determine how far the magnetism of Mesmerism was analogous to that to be employed in telegraphs."

A reporter sought out Morse in the gallery. He was not amused.

"I have an awful headache," said Morse, holding his head.

"You are anxious?"

"I have reason to be. . . . I have spent seven years in perfecting this invention. . . . If it succeeds I am a made man. If it fails I am ruined."

After several readings and after it was shorn of the Mesmerism and Millerite amendment (twenty-two congressmen had voted for it!), the House voted for the telegraph 89 to 83. No southerner voted for it.

On the last day of its session, the Senate still had some one hundred and forty bills to pass. When the lamps were lit at nightfall and his bill had still not been reached, Morse left and went to his cheap hotel room, sure that his cause was lost. After paying his hotel bill he would have just enough for his train fare plus thirty-seven and a half cents. He went to bed.

But he'd won. President Tyler signed the bill at midnight.

Next morning Annie Ellsworth, the daughter of a friend, brought Morse the tremendous news. Beside himself with joy, Morse promised her that she could send the first

Preceding pages: Morse's first telegraph apparatus (left), and Morse in his workshop, about 1836. Opposite top: Civil War tacticians depended on the telegraph for controlling troop movements, supplies, and railroad operations. Bottom: Morse's Louvre painting. Left: Morse in old age and Morse receiver.

message over the electric telegraph. She chose, "What hath God wrought?"

The test line was to run forty miles between Baltimore and Washington, along the right of way of the infant Baltimore & Ohio railroad. Tricky "Fog" Smith, maneuvering as usual, managed to emerge as construction contractor. Morse and Vail decided to put the wires underground in lead pipes. Trenching was needed and Ezra Cornell, the construction engineer, designed and built a trick plow which dug a trench, buried the lead pipe with the wire inside, and then neatly covered it over with dirt. There were troubles. The contractor who was to provide the lead pipe with the wire inside was very late delivering it, and a new contractor had to be hired. After nine miles were laid it was discovered that the insulation was defective. "Fog" Smith was playing crooked games with subcontractors, was in fact stealing and swindling.

The underground operation was suspended. Morse decided to use poles instead. But not until $23,000 of the $30,000 appropriation had been spent. The poles went up fast. Two hundred feet apart, of chestnut, unbarked, twenty-four feet tall, using broken bottle necks as insulators, they followed the railroad tracks as such poles and wires have ever since.

On May 1, 1844, the Whig convention opened in Baltimore. By then the telegraph had crept within twenty-two miles of Baltimore at Annapolis Junction. Morse sat near his instrument in the capital. Vail was at the other end of the wire. As the afternoon train from Baltimore reached the Junction delegates yelled, "Three cheers for Henry Clay, three cheers for Frelinghuysen!" Vail clicked off the news on his key. Morse in Washington watched his tape and announced, "The ticket is Clay and Freylingmusen." The political hangers-on loafing in Morse's room jeered, "Anyone could guess Clay would head the ticket, but Freylingmusen! Who the hell is Freylingmusen?"

They found out about an hour later when the B & O train pulled into Washington. But mostly they found out that the telegraph worked, and that it was faster than the fastest train.

It was not until May 24 that the telegraph line was finished, and Annie Ellsworth's Biblical quotation went from Washington to Baltimore over Mr. Morse's electromagnetic telegraph. The next day the Democrats convened to nominate a presidential candidate. Morse, the future Copperhead, was happy to report that Van Buren, who was against the spread of slavery into Texas, couldn't get a two-thirds vote and the nomination. Then as politicians crowded around, Morse's telegraph register clicked out: "Polk is unanimously nom. 3 cheers were given in convention."

There were many more than three cheers for the Great Electric Telegraph. Suddenly, from a ridiculous toy laughed at by southern congressmen, it became the wonder of the age, the proof of American genius. People came to find out how much it would cost to send parcels to Baltimore via telegraph. They asked for news of the Philadelphia anti-Catholic riots, for news of the 1844 election. A family heard that a relative in Baltimore had been murdered. The telegraph almost instantly told them he was alive and well. And the story about a couple of fellows playing chess by telegraph was as exciting as seeing astronauts in a space capsule shaving for the television cameras.

Actually, neither Morse nor the United States government really knew what to do with the telegraph. At first Morse hoped the government would take it over and pay him as the French government had paid Daguerre. In the end, as the railroads had been, and as the telephone, the electric power system, radio, and television would be, the telegraph became a bonanza for speculators. First Morse, "Fog" Smith, and a man called Kendall formed the Magnetic Telegraph Company from which Morse managed to shut out Vail and many of his early associates. Then a promoter named

Right: Alexander Graham Bell as a youth. Opposite: Page from a treatise on "Visible Speech" by Alexander Melville Bell, father of the inventor of the telephone. Each letter of the Visible Speech alphabet shows how to arrange the tongue, lips, and throat to make the specialized sounds required to speak any language.

O'Rielly got into the act to form stock companies which strung thousands of miles of wire in every state east of the Mississippi by 1848.

Patent battles over claims by rival inventors, fights with partners, angry quarrels with Professor Henry and with Doctor Jackson, who claimed that he had been inspired to conceive the telegraph aboard the *Sully,* took much of the sweetness out of Morse's triumphs and profits.

By 1856 Western Union was organized and the pressure slackened. Morse could have settled down to enjoy his fame and his wealth. Instead, he reverted to his old militant Know-Nothingism. After the fall of Fort Sumter he became president of the Society for the Diffusion of Political Knowledge, an anti-Lincoln, proslavery outfit. In 1864 he campaigned for McClellan against Lincoln.

On April 2, 1872, telegraphers in every country tapped out the dots and dashes signaling Morse's death.

The Talking Wire

"If Bell had known more about electricity," said Moses G. Farmer, "he never would have invented the telephone." Mr. Farmer, himself a noted inventor and the designer of Boston's fire-alarm system in the 1850s, was referring to the fact that among electrical experts of the 1860s and seventies conventional wisdom would not have accepted the possibility of a telephone resulting from Alexander Graham Bell's line of reasoning.

Clever Mr. Farmer might better have said: "If Bell had known less about the human ear, he never would have invented the telephone."

For even in his twenties, young Aleck Bell knew as much about speech, acoustics, and the peculiar characteristics of the human eardrum as anyone in the world. And it was in proceeding from this knowledge that he eventually achieved the telephone. (The word itself, a compound of Greek roots meaning "far-off voice," evidently was first used by Sir Charles Wheatstone, who applied it to his Magic Lyre, a nonelectrical sound transmitter he developed experimentally in 1819.)

Alexander Graham Bell was born a Scotsman in Edinburgh on March 3, 1847. His father and grandfather were philologists, and exponents of that important and respected nineteenth-century art, elocution—public speaking with appropriate gestures. His father, Alexander Melville Bell, invented a complex phonetic system called "Visible Speech." It involved charts of written symbols which represented sounds—a new kind of universal alphabet in which each letter showed how to arrange the tongue, lips, and throat to make the specialized noises necessary to speak, say, Urdu or Choctaw or Yiddish, or even correct English, for that matter. Bernard Shaw in *Pygmalion* used the system to have Henry Higgins transform Eliza Doolittle's Cockney nasalities into the round tones of the Queen's English.

Aleck Bell grew up in Edinburgh among the vibrations of noble sounds, correctly modulated voices, music, song. He became a first-rate pianist and had visions of playing on the concert stage. But he became more interested in dabbling in scientific matters than in the necessary long hours of practice. It is worth noting that after four years of high school in Edinburgh he had little further formal education. In 1862, when he was fifteen, he was shipped off to join his grandfather in London. And like Henry Higgins, Grandpa Bell set himself the task of reshaping Aleck into a proper London gentleman, complete with a decent education, by forcing him to plow through his formidable library and giving him the same tough elocution lessons he gave to Londoners for pay. It was high pay; thirty to one hundred guineas for a three-month course. Years later Aleck Bell wrote that the time with Grandfather Bell "converted me from a boy somewhat prematurely into a man."

In 1863, when Aleck Bell's father came to London to retrieve his son, they stopped in to see Sir Charles Wheatstone in order to inspect his "speaking machine," a mechanical contraption for imitating the human voice, which Sir Charles had built twenty-five years earlier, when he was only twenty-

ORGANIC FORMATION OF THE PRINCIPAL ELEMENTS OF SPEECH.

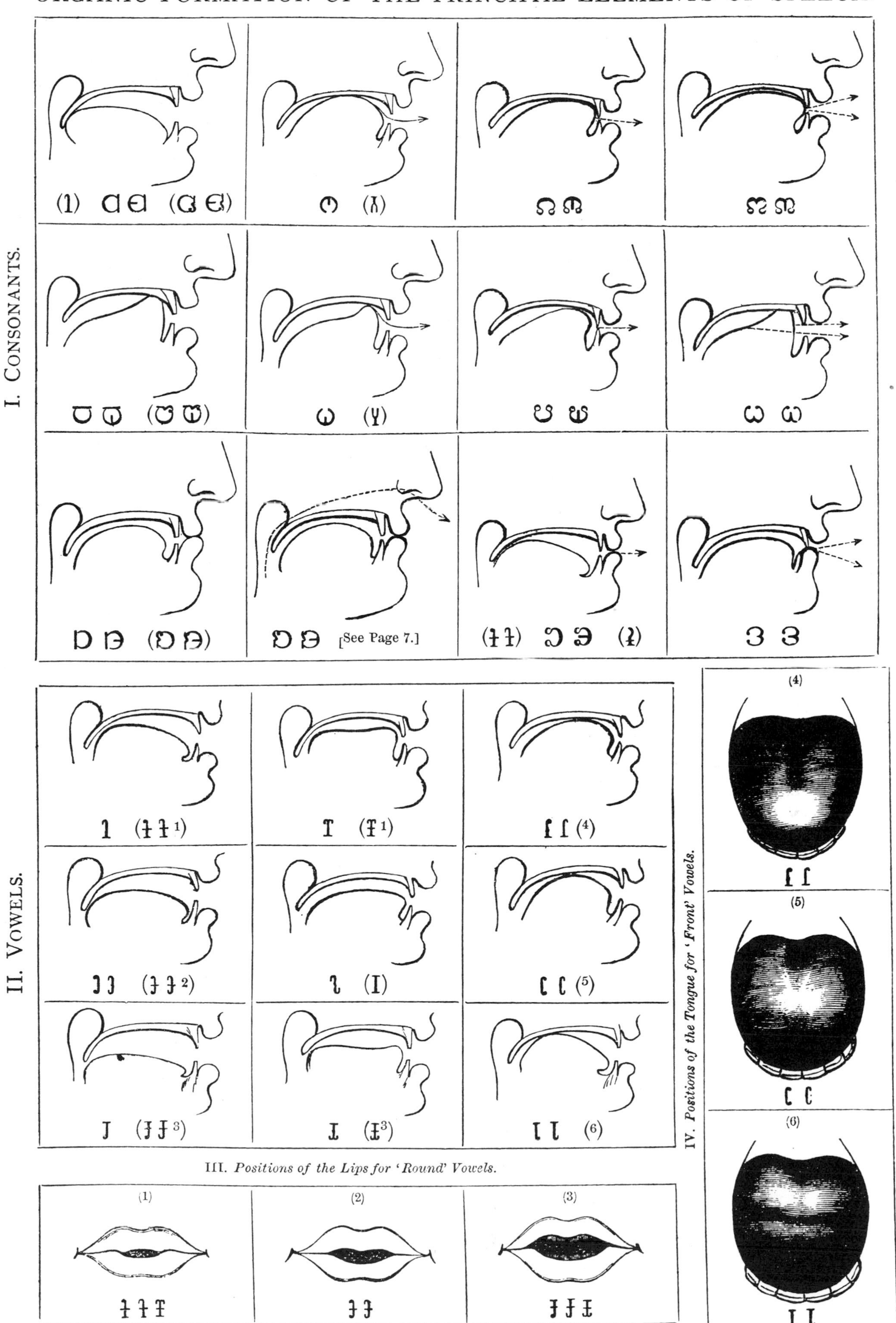

Opposite, top: Bell's attic laboratory at 109 Court Street, Boston, as reconstructed from original bits and pieces after it was torn down in the 1920s. Note wet-cell batteries at right end of bench. Bottom: Bell's magnetic telephone. Left: Bell's "gallows-frame" transmitter.

one. At the same time they must have seen Wheatstone's Enchanted Lyre, which Wheatstone had made even earlier. This device used tuned metal rods which vibrated in sympathy with musical notes sent from a distance through a solid conductor. It is not too unlikely that this long-distance (albeit nonelectrical) transmission of sound planted a seed which a dozen years later grew into the telephone. In any case, when Aleck Bell returned to Edinburgh he and his brother Melly, encouraged by their father, built from rubber, scraps of wood, and cardboard a speaking machine of their own. It sounded like a barely intelligible Donald Duck with laryngitis.

In August, Weston House, a boarding school in Elgin, Scotland, hired young Bell, then all of sixteen, for a year as a teacher of music and elocution. While performing his duties at Elgin he continued his experiments in speech and acoustics. Among other things he discovered a way of finding the exact pitch of vowels by means of tuning forks —experiments which were another step on his way to the telephone.

He wrote of his experiments to his father's friend, the famed and eccentric phonetician Alexander Ellis. Ellis dampened Aleck's visions of instant fame for his discoveries by informing him that the great German scientist Helmholtz had conducted similar experiments. Moreover, using electromagnets he had been able to synthesize vowel sounds by vibrating tuning forks electrically. Bell was unable to read Helmholtz's German account, which Ellis sent him. Depending on Ellis's none-too-good translation, he got the mistaken impression that Helmholtz's rheotome (as it was later known) transmitted sound by wire. He got the idea that speech, too, might be sent over a wire. But as yet he had not the vaguest notion that he might be the man who would do it.

In 1866 Aleck Bell went off to teach at Somersetshire College in Bath. While there he and a couple of friends amused themselves by setting up a pair of old Wheatstone needle-telegraph instruments and signaling each other through wires hung on the outside walls of several houses. Now he learned something about batteries, circuits, and electromagnets.

Meanwhile, Bell senior traveled to the United States and Canada for a lecture tour promoting his book "Visible Speech." While his father was abroad, Aleck managed his affairs in London. He taught Visible Speech and elocution and found particular satisfaction in teaching deaf children to speak. But hard work, English weather, and the dank and smoky fogs of Victorian London began to affect his health. By the time his father returned, Aleck was near a breakdown. As two other sons already had died, the Bells decided they had had enough. In 1870 they headed for the New World. The telephone would be an American invention.

The family settled in Canada, in the town of Brantford, Ontario. While Melville Bell lectured and taught, Aleck recuperated and read Helmholtz's *On the Sensations of Tone.* When Melville was offered a teaching job at the Boston School for the Deaf, he refused it and young Alexander went instead. His success in teaching the school's deaf children was sensational. To pad out his small salary he also gave private lessons to correct stammering and speech defects among the children of wealthy parents.

In 1872 tuning forks again began to twang inside Bell's head. He started working on his Harmonic Multiple Telegraph, a means of sending several messages simultaneously over one wire. Western Union would pay a fortune for such an invention. His idea was based on Helmholtz's rheotome. If the electrically induced vibrations of a tuning fork could be sent over a wire, an electromagnetically controlled tuning fork at the other end of the wire would vibrate at the same frequency. If, Bell reasoned, he were to send several such different electrical notes over the same wire at the same time, and there were several different electromagnetically excited tuning forks at the receiving end, each sending fork and each receiving fork would vibrate at its own frequency and, therefore, send its own separate message over the same wire at the same time. As many messages could be sent simultaneously as there were notes on the musical scale.

Despite his lack of a degree—which did not trouble academics in those days—Boston University offered Bell a professorship of Vocal Physiology and Elocution, a crucial opportunity for him. He now had a platform for making speeches to important groups, "the cream of Boston's intellec-

Right: Bell's "Centennial" transmitter. Opposite: Bell's first sketch of his telephone. Note that the little men are not merely talking— they're yelling!

tual men." One lecture, sponsored by Massachusetts Institute of Technology's Society of Arts and Sciences, was addressed to some five hundred people, including "the finest minds of Boston." Now, too, he gained access to MIT, which meant permission to experiment with Helmholtz's apparatus and with other devices of the time which bore on study of the voice. And he carried on his own work.

Late in 1873 he started to substitute "reeds" of brass or steel (like the reeds in musical instruments) for tuning forks. Each was mounted so it projected like a tiny diving board over a small cup of mercury. He fastened a platinum wire at the free end to make electrical contact with the mercury. To vibrate the reed mechanically he blew down on it through a cone. To change the reed's pitch he altered the length of the projection by sliding the fixed end under a screw clamp. Each reed thus was tunable. A tuning fork wasn't.

Bell then wired two of his reed rheotome transmitters, each of a different pitch, in parallel with each other and in series with two steel-reed receivers. He had tuned each receiver to a matching transmitter. But he ran into trouble. One transmitter made its similarly tuned receiver sound. The other one didn't. That was what he wanted. But when he tried both at once, neither sounded, except that he could feel both of them vibrating slightly. Bell blamed the balkiness on the reed of his second receiver, assuming it was improperly tuned. While diddling with it, he put his ear against the balky reed. It emitted a faint hum. Bell didn't realize it, but what he heard was the infant telephone, making its first weak birth cry.

Bell next immersed himself in trying to work out circuits. He floundered and lost himself in a maze of magnets, wires, and batteries. Then he decided that he needed a new kind of transmitter without the fussy and unreliable mercury-cup switch. In a flash of pure genius which in the end led to his great invention, he decided to scrap his transmitter entirely. This meant using his receiver for both transmitting *and* receiving. He reasoned that if the receiver's reed were vibrated—an air current could do it—its movement would, by induction, cause the current in the coil of its electromagnet to fluctuate. It would be a *continuous* current, not an intermittent one. The second receiver's reed would then vibrate in tune with the first one.

Bell still had no inkling that he almost had the secret of the telephone in his grasp. He assumed that the faint sound of the receiver's reed would be too weak for the human ear to detect. He didn't even bother to listen. Still dissatisfied, he went on to try other ways of making his Harmonic Telegraph work.

Bell was busy with other things during the fall and winter of 1873-74: Boston University classes, "Visible Speech," and private pupils—especially one, a bright and pretty girl who had been totally deaf from scarlet fever since she was five years old and as a result had also lost much of her ability to talk. Her name was Mabel Hubbard. She was not quite sixteen years old and she was the daughter of Gardiner Greene Hubbard, a lawyer, a financial wheeler-dealer, a street-railway tycoon, an all-round promoter. Almost as much as Bell he would make the telephone system a reality.

At first Mabel did not like her young therapist. She was used "to the dainty neatness of Harvard students." And Bell, tall and darkly intense, "dressed badly and carelessly." Still, she found him to be a wonderful teacher. As for Bell, he was much smitten with Mabel Hubbard and took every opportunity to visit the family's garishly Victorian house in Cambridge. There he told Hubbard about his Harmonic Telegraph. Another man might have shown polite interest, but Hubbard was elated and excited. So much so that he seemed about to kiss poor startled Bell right there in the middle of the red velvet and golden oak drawing room. For Hubbard was engaged in a battle with Western Union which, he claimed, was retarding the country's economy by monopolizing the telegraph. He charged that another company, using some form of multiple telegraph, could lower rates and make telegraphy almost as cheap as letter mail. The new company chartered by Congress would be called the United States Postal Telegraph Company, since the Post Office would handle the messages. And Hubbard would be one of the chief officers. Hubbard had a sharp eye for a buck.

Another rich parent of a Bell pupil, Thomas Sanders, heard that Bell had told Hubbard about his Harmonic Telegraph and became suspicious. He knew about Hubbard's

Boston, Mass. To Miss Frances B. Symonds
Aug. 21st 1876 from A. G. B.

As far as I can remember these are the first dra[wings]
of my Telephone – or "instrument for the [trans-]
[m]ission of vocal utterances by telegraph."
A. Graham Be[ll]

BOSTON
SALEM
THE TELEPHON

Opposite: In 1877, Bell lectured about his telephone in Salem, Massachusetts. He carried on a conversation with his assistant Thomas A. Watson, eighteen miles away in Boston, and further regaled his audience by singing "Yankee Doodle Dandy" over the telephone. He was a smash hit.

sometimes tricky financial dealings and warned Bell to cover his invention immediately with a patent. Bell didn't have the money to build patent models and pay legal fees, so a three-way deal was made. Bell, Hubbard, and Sanders each were to own a third of the invention. Sanders and Hubbard would put up money, Bell his ideas. Bell rushed into the pact partly because he had heard that the electrical engineer Elisha Gray seemed to be working along similar lines. The three-way partnership was destined to become the biggest monopoly in all history: the American Telephone and Telegraph Company.

Although Bell had interested Hubbard and Sanders only in his Harmonic Telegraph, another idea had implanted itself in his mind before his meeting with Hubbard in the fall of 1874. During the previous summer, while vacationing at his father's home in Canada, his thoughts wandered to the idea that a vibrating reed causing an undulating current in the coil of an electromagnet might transmit the sound of a voice.

He experimented with the human ear—a real human ear from a cadaver, complete with drum and its attached bones. To one of the bones he attached a thin straw which acted as a stylus to visibly trace the sound waves on small sheets of smoked glass which he carefully pulled past the straw stylus as he yelled into the dead man's ear. He was amazed at the power of the ear drum to move the heavy bones and through them convey complicated sounds to the tiny point of the straw. He made other experiments and sketched a "harp apparatus" using a series of reeds each only very slightly different in pitch, which he thought might transmit speech. After returning to Boston he wrote his parents about his outlandish ideas for talking over a wire: "I have scarce dared to breathe to anybody for fear of being thought insane."

Early in 1875 Bell started working with a young assistant who would play an important role in the invention of the telephone. His name was Thomas A. Watson. Watson worked for a man named Williams who owned a workshop where he built electrical apparatus in small quantities: fire alarms, telegraphic devices, electric bells, batteries, and the like. He also invited headaches by constructing gadgetry for a stream of sometimes rather odd inventors. Bell had increased his use of the Williams shop to build his apparatus after Hubbard and Sanders had begun to invest in his Harmonic Telegraph. And soon, because they seemed to hit it off so well, Williams assigned Watson to Bell's work.

"No finer influence than Graham Bell ever came into my life," wrote Watson in the 1920s: "His punctilious courtesy to everyone . . . was a revelation." Watson was impressed by Bell's table manners (his use of a fork instead of a knife for putting food in his mouth), and tried to copy them. Bell talked to Watson of books and authors—Helmholtz, Gauss, Huxley, and others.

Bell diffidently mentioned his ideas about a telephone to Hubbard. Hubbard wasn't impressed. He wanted something solid, something that would make money now! He told Bell to stick to the Harmonic Telegraph. And that's what Bell and Watson did during that winter of 1875. By late February they had built a model good enough to patent. When Bell got to Washington, however, he was appalled to find that Elisha Gray had patented a sort of Harmonic Telegraph only two days before. What Bell did not know was that the Harmonic Telegraph—neither his nor Gray's—would ever amount to much anyway. Edison's Quadraplex Telegraph would be the successful design—Quadraplex meaning that four messages could be sent over one wire.

Before leaving Washington, Bell went to the Smithsonian Institution to see its director, the prestigious scientist Joseph Henry, who had done so much to help that ungracious inventor of the telegraph, Samuel Morse. He told Henry about his ideas for transmitting speech by means of his "harp" apparatus and also about his ideas for a telephone. Henry thought that he had got hold of "the germ of a great invention."

"What would you advise me to do?" asked Bell. He wondered whether he ought to publish his ideas for others to carry out, or struggle to make them work himself. Henry hadn't forgotten that he had had basic ideas for the telegraph which he squandered on Morse. He told Bell to work out the idea of the telephone for himself.

Bell pointed out that he simply hadn't enough knowledge of electricity. "Get it!" snapped Henry.

That was easy enough for Henry to say. But Bell was at that time perhaps the busiest young man in the Western

WESTERN UNION TELEGRAPH CO.
JEWELRY
179
PHOTO
S.F. MYERS & CO
LITH
WATCHES
JEWELERS
CO.
LOCKWOOD
PETERSON & ROYCE DIAMONDS
189
PETERSON & ROYCE DIAMONDS
WATCHES DIAMONDS JEWELRY 203
BIGLEY & CONWAY MERCHANT TAILORS
S.F. MYERS & CO WATCHES 179
DUNLAP & CO. 181
ROGERS PEET & CO CLOTHIERS
HENRY L. SANDERS MERCHANT TAILOR 175
205 BROADWAY DELIVERY WAGON
BROADWAY
BROADWAY
MARVIN SAFE CO.
MARVIN SAFE CO 265 BROADWAY.
KRADLER MANUFACTURER OF GOLD RINGS & LOCKETS 179 BROADWAY.
SOLID GOLD RINGS
FACTORY 185 EDDY ST
HUDSON AND FARNUM
196
THOS. F. ARNOLD SUCCESSOR TO ARNOLD & WEBSTER BAND RINGS
194
J.W. RICHARDSON & CO.
DENNISON
ALFRED H. SMITH & CO. DIAMONDS 33 HOLBORN VIADUCT LONDON.
23 ST 9 AV.
SOUTH FERRY
BROAD WAY.
No1 COURTLAND ST OPP. MAIDEN
TRUNKS

Opposite: Scene in lower Manhattan in the 1880s shows complex spider webs of telephone and telegraph wires on poles. A tangle of fallen wires, buried under deep snow after the blizzard of 1888, caused a frightful mess, which led to the later requirement that wires go underground. Below: British advertisement for "Articulating or Speaking Telephone."

THE TELEPHONE.

The Articulating or Speaking Telephone of Professor Alexander Graham Bell has now reached a point of simplicity, perfection, and reliability such as give it undoubted pre-eminence over all other means for telegraphic communication. Its employment necessitates no skilled labour, no technical education, and no special attention on the part of any one individual. Persons using it can converse miles apart, in precisely the same manner as though they were in the same room. It needs but a wire between the two points of communication, though ten or twenty miles apart, with a Telephone or a pair of Telephones—one to receive, the other to transmit, the sound of the voice—to hold communication in any language. It conveys the quality of the voice so that the person speaking can be recognised at the other end of the line. It can be used for any purpose and in any position—for mines, marine exploration, military evolutions, and numerous other purposes other than the hitherto recognised field for Telegraphy; between the manufacturer's office and his factory; between all large commercial houses and their branches; between central and branch banks; in ship-building yards, and factories of every description; in fact wherever conversation is required between the principal and

Opposite: In early telephone exchanges, the operators were boys. They were soon replaced by young ladies, who were more polite to the customers. The boys had tended to tell off irate subscribers.

Hemisphere: He was teaching classes of deaf children, lecturing to teachers who would put into practice his father's "Visible Speech" system, working on ramifications of the Harmonic Telegraph, including a complicated and ultimately unworkable "autographic telegraph," and trying to court Mabel Hubbard.

In June of 1875, nonetheless, on one of those miserably hot and humid Eastern seaboard days, Bell, sweating in the attic of the Williams shop, at last stumbled upon the means that would make the telephone work. He and Watson were in separate rooms trying to tune three transmitters to pitches different enough from each other so that each would excite only its own reed receivers. One of each of the receivers was in Bell's room, the others in Watson's room. Bell tried the first transmitter and its receivers sounded properly. The second transmitter also caused its receivers to act correctly. When Bell sounded the third transmitter, Bell's receiver worked but Watson's did not. Bell yelled to Watson to pluck at the inoperative reed on his receiver to free it up. Reeds sometimes stuck. Bell disconnected the transmitters and the batteries from the circuit while waiting for Watson to twang his recalcitrant reed. Meanwhile he kept watching its duplicate in his room.

And then, unbelievably, as Watson twanged his reed, Bell's reed vibrated in unison and faintly, very faintly, gave off the same twanging sound. Bell became excited. He knew instantly what had happened. The vibration of Watson's reed had generated current—an undulatory current—by induction alone. And Bell knew then that the sound of that reed was a complex of audible frequencies, as complex as the sound of a voice. He knew he had the telephone!

It was the undulatory current which made the telephone possible. In telegraphy each signal—either a dot or a dash—was a sharply defined pulse of electricity. Each pulse had the same intensity. In telephony a continuous current, whose intensity could be varied as sound waves were varied, was necessary.

Bell decided that diaphragms were the answer. He roughly drew a sketch of such a transmitter and asked Watson to make a couple of them. A few days later Watson brought in the new instruments. Their chief components were a U-shaped electromagnet and a diaphragm made of stretched goldbeater's skin—a thin membrane made from the large intestine of an ox and used in the making of gold leaf. A steel-reed armature was hinged from one pole of the electromagnet. The other end of the armature extended over the other pole and was attached to the center of the diaphragm.

In the attic Bell yelled into one of the new transmitters. Watson on the floor below wanted desperately to hear something in his reed receiver and said he heard a very weak "something." When Watson did the shouting into the transmitter Bell could not hear him. Anyhow, the ox membrane proved too fragile for further tests. First the reed tore away, and when it was refastened more strongly the membrane split across.

About a month later, using a stronger membrane and a more delicate armature, Bell bellowed song and poetry into the newly revised transmitter. Watson charged up the stairs yelling, "I could hear your voice plainly. I could almost hear what you said!" Oddly, although the speech transmitted by this first of Bell's telephones was unintelligible, it was basically the instrument patented ten months later.

Why the delay?

First, Gardiner Hubbard, still without confidence in the idea that it was possible to talk over a wire, bullied poor Bell into getting back to work on the harmonic and autographic telegraphs. Then Watson got sick. Then Bell became involved in complicated and (to this writer) utterly boring negotiations with the Hubbards over whether or not he might marry their precious daughter. Getting together specifications for a patent took up more time. Further, Bell held back on his American patent, hoping that a Canadian newspaper publisher, George Brown, would buy the foreign rights. (Bell's deal with his partners covered only the American rights.) Bell had asked Brown for $50 a month for six months plus the legal fees for filing the patent. Brown thought so little of "undulatory current" and the telephone that he let the whole business lapse without bothering to tell Bell.

Hubbard, although convinced that the telephone was less important than the Harmonic Telegraph, finally be-

came impatient and, without telling Bell, filed an application for a patent on the telephone on the morning of February 14, 1876. Incredibly, two hours later Elisha Gray filed a caveat (an intention to patent) for a not too dissimilar telephone idea.

Bell, himself the most honest of men, couldn't believe that Gray had somehow got wind of the telephone and was trying to steal his idea. And perhaps he wasn't. For it is not unusual for men working in the same field to be struck by the same idea at about the same time—a time when related techniques not only make an idea workable but also when the world is ready to use it. Similarly, ten years later, Daimler and Benz, unknown to each other, developed the automobile almost simultaneously.

The basic patent on the telephone, No. 174,465, was granted to Bell on March 7, 1876. It was the most valuable patent ever issued. Bell was only twenty-nine years old.

Patent or no, the telephone still did not speak. It mumbled and muttered. Bell and Watson intensified their experiments. The telephone first spoke three days after the patent for it was issued. Bell had Watson make a device in which a platinum needle dipped downward from a membranous diaphragm into a dish containing water plus a little sulfuric acid. The idea was that the electric current would vary as the needle rose and fell as the diaphragm vibrated.

Watson was in Bell's bedroom with a reed receiver against his ear. Bell was down the hall in his laboratory. There were two closed doors between them. As Bell later wrote in his notebook: "I then shouted into the mouthpiece the following sentence: 'Mr. Watson—come here—I want to see you!' To my delight he came and declared that he had heard and understood what I said."

Fifty years later, in his autobiography, Watson described the now-famous sentence as being a call for help because Bell had spilled some of the acid on his trousers. But Watson must have mixed up the incident with another one. Further, Watson's version—the one that went down into the schoolbooks—leaves out the words "to see."

On March 27, with his father watching him as he fiddled with various combinations of receivers, diaphragms, and armatures, Bell tried using a magneto telephone with a small, flat-coiled clock spring as the transmitter diaphragm armature. Now the magneto telephone spoke. It said, "Papa." The magneto telephone would be the basis for soon-to-be commercial telephony.

The great event of 1876 was the Centennial Exhibition in Philadelphia. Bell, frantically busy with his teaching commitments, almost didn't get there, but Gardiner Hubbard and Aleck's wife-to-be, Mabel, saw to it that he showed up. Elisha Gray was already there with his telegraphic gadgetry when Bell started setting up and repairing his equipment, some of which had been damaged en route.

The head of the committee of judges in the electrical department was the famous British physicist Sir William Thomson (later Lord Kelvin). Bell had told him about his telephone and Thomson was eager to examine it. But Thomson spent so much time looking at other exhibits that he asked if he and the committee and a special guest, Dom Pedro, Emperor of Brazil (who had visited the School for the Deaf and met Bell in Boston a week earlier) might examine Bell's devices on the following Sunday—June 25. This day was to be remarkable for another historical event: the destruction of General George Custer and his troopers at the Little Big Horn.

The public was barred on Sundays. The judges and the emperor and his party had the huge, hot, glass building almost to themselves. First they listened (it seemed forever to Bell) to Elisha Gray, who was exhibiting his telegraphic novelties under the aegis of Western Union. The judges, broiling under the sun streaming through the acres of glass, were ready to quit right there. But, as Bell proudly told it to his grandchildren years later, Dom Pedro insisted upon examining Bell's exhibit. There he and the others saw Bell's Harmonic Telegraph and other telegraphic gimcrackery.

But there was also a telephone. And a new kind of receiver with a cylindrical magnet. Later it was to be known as the "Centennial iron-box receiver." Wires from the new receiver ran a hundred yards or more to Bell's three transmitters at a far end of the building. One was of the variable resistance type with a thin rod hanging down from a membrane into a cup of water and acid. The other two were electromagnetic.

After explaining his Harmonic Telegraph, Bell

Right: After his retirement, Bell experimented with aircraft and the first hydrofoil. Here the HD-4 rises on its foils on Baddeck Bay, Nova Scotia. In 1919 it set the world record of 70.66 mph. Opposite: Alexander Graham Bell opens the transcontinental long-distance line in 1912.

rushed to where his transmitters lay and started singing into one of them. Sir William, with an iron-box receiver against his ear, froze into startled immobility at the sound of ghostly singing. Then, thinly, weirdly, the words "Do-you-understand-what-I-say?" came through. "Yes!" he yelled, losing his British cool. "Where is Mr. Bell? I must see Mr. Bell!" An assistant started running to get Bell, but Thomson went by him like an Olympic sprinter and reached Bell as he repeated "Do you understand what I say?" into the mouthpiece of the transmitter. Out of breath, Sir William puffed, "I heard the words 'what I say' " and begged for a further demonstration.

It was the volatile Dom Pedro's tremendous reaction that most pleased Bell. He not only listened but insisted upon talking into the transmitter to one of his party at the other end of the wire. "My God," the emperor roared, "it speaks Portuguese!"

Even Elisha Gray expressed his delight over the telephone demonstration. Further, he made no claim that he had been first to invent the telephone. That came later.

Not everyone was enthusiastic about the telephone. When a Professor Tait in Edinburgh heard about it through Sir William, he said, "It is all humbug, for such a discovery is physically impossible."

Bell continued to improve his invention. He glued a large disc of thin steel to the membrane, then he used only the steel disc without the membrane on the transmitter. He tried new types of magnets. Not only did the telephone speak more clearly, its range increased.

In 1877 in Salem, Massachusetts, Bell lectured on his telephone to an audience packed into Lyceum Hall. He not only carried on a conversation with Watson in Boston, eighteen miles away, but regaled his audience with renditions of "Auld Lang Syne" and "Yankee Doodle." The press, not only in America but in London and Paris, began to take notice, which, oddly, it had not done before.

Bell repeated his successful Salem lecture in New York, Boston, Providence, New Haven, and elsewhere. Opera singers, quartets, the Boston Cadet Band, even Watson performed. Bell's mother-in-law-to-be made pointed allusions to P. T. Barnum.

In July Bell at last married Mabel. In August Mr. and Mrs. Alexander Graham Bell sailed for England. There he was lionized, gave lectures, and demonstrated his telephone to the queen, causing some consternation by a grave infraction of court etiquette. When she turned away for a moment, Bell dared to touch her hand and offer her the receiver. The Bells stayed abroad for a year.

Now came the bad part, the business part, the fight for foreign patents, the raising of money, the corporate infighting, and the battle against other claimants to the invention. Bell, no businessman, hated it. Worst of all was the attack on Bell by Western Union. Led by its long-bearded president, William Orton, Western Union marshalled Gray, Thomas Edison, who had designed an improved transmitter, and others to break Bell's patent. The giant telegraph combine even set up a rival company, the American Speaking Telephone Company, which, when the Bell Company started operations, reduced rates and cut Bell wires.

Edison's carbon transmitter was better. But the Bell Company found an even better one, designed by Emile Berliner, a German immigrant who had filed a caveat thirteen days before Edison applied for a patent. Watson hired him.

In 1879 Western Union and the Bell Company came to terms. W.U. agreed to assign all of its patent rights for twenty per cent of telephone rental receipts for the next seventeen years. But the patent wars were not over. For eighteen years the Bell Company fought some six hundred cases brought by claimants who included a fair number of fools and crooks. Alexander Bell spent untold hours on the witness stand.

One of the more bizarre conspiracies, during Cleveland's administration, involved several senators, an ex-governor of Tennessee, and one Augustus H. Garland, Attorney General of the United States. This crew tried to use the power of the United States Government to break the Bell patent in favor of a paper corporation they controlled. They didn't succeed.

After 1881, Bell left the improvement of the telephone to others. He was a rich man and had forty years of his life ahead of him to do exactly as he wished. In spite of

Right: Bell and the Wright brothers. Orville (near limousine door), Bell, Wilbur, and a factotum of the Smithsonian Institution leave the Institution after the Wrights' receipt of the Langley Medal for aviation achievements in 1910. Opposite: Heinrich Hertz, whose experiments with what came to be known as Hertzian waves led to radio and television.

sleeping past noon he got a lot done. It must be admitted that he stayed up very late working on many great projects. Among them was an electrical apparatus he called an "induction balance"—a metal detector—which was used in a vain effort to locate the assassin's bullet in President Garfield.

Unlike other successful inventors who turned out badly—Morse, for example—Bell was a fine man, who worked for the betterment of the human condition. He organized greater means for teaching the deaf. Helen Keller dedicated her autobiography to him: "He is never quite so happy as when he has a little deaf child in his arms."

Bell financed scientists and the first decade of that still highly respected journal, *Science.* His wide-ranging interests included aviation (he financed Glenn Curtiss and built tetrahedral-winged aircraft), marine navigation (his were the first successful hydrofoils), sheep raising, and work on an early iron lung. He invented the photophone, a method of speaking over a beam of light which though not a commercial success, he considered his most important invention. The list goes on and on.

But Bell never succeeded in convincing anyone that they ought not to answer the telephone by saying, "Hello." He thought that the correct salutation was "Hoy! Hoy!"

He died a naturalized American citizen on August 2, 1922, in Nova Scotia, where he had for many years enjoyed his happy experiments.

Sound Without Wires

Young Guglielmo Marconi was by no means the first to realize that electromagnetic waves from the violently crashing sparks generated by an induction coil could be detected at a distance. But he was first to put together the parts of existing theory and hardware in such a way as to send and receive dot-dash messages.

In 1842 Professor Joseph Henry, the electrical genius who had helped Morse with his telegraph and Bell with his telephone, showed that the effects of an electric spark could be detected from a distance of about thirty feet. Twenty-five years later Professor James Clerk Maxwell of the University of Edinburgh propounded the wave theory of light, heat, and electromagnetics, and described the waves that made wireless telegraphy, radio, and television possible.

It remained for Heinrich Hertz, in Karlsruhe, Germany, in 1886 to demonstrate that it was practicable to create and detect the waves of electrical radiation. Hertz—whose name survives today as the designation for a one cycle/second unit of frequency—devised apparatus to reflect and focus the newly discovered waves. His electric-wave generator consisted of an induction coil feeding a spark gap from which a pair of plate-like conductors extended. The conductors corresponded to what became the antenna and ground wires of later radio transmitters. Hertz's receiver was a wire ring with a small opening in its circumference. When electromagnetic waves arrived, minuscule sparks passed across the opening. With this primitive loop antenna Hertz was not only able to prove that he could receive emissions from his spark transmitter, but was also able to measure their strength and determine their direction.

Many other savants took up where Hertz left off; among them, the Frenchman Edouard Branly and the Italian Augusto Righi had most to do with inspiring Marconi. Branly was a physicist, a professor at the Institut Catholique in Paris. Without his detector of radio waves, later named the "coherer" by the great British investigator of electromagnetic phenomena, Sir Oliver Lodge, Marconi could not have put together his first wireless-telegraph apparatus. Branly happened upon the coherer in 1885, while he was experimenting with theories relating to the nervous system. He found that nerve fibers are not continuous, but are chains of neurons which, although close to each other, are not in contact. He built an electrical gadget approximating how he thought neurons worked. It consisted of a glass tube almost filled with loose iron filings, a galvanometer, and a battery, wired together in a closed circuit.

Hertzian—electromagnetic—waves generated

twenty-five meters away caused the iron filings to cohere, while the galvanometer showed the effect on the battery current. He also discovered that the metal filings could be discohered by a slight concussion. Branly was elected to the French Academy in 1891, when he demonstrated his device to that august body. Lodge improved the coherer by incorporating a mechanical tapper and other improvements which increased its sensitivity. He exhibited his coherer to the British Association at Oxford in 1894 and received signals across a 150-yard distance but, strangely, didn't realize that he had an instrument which might be used for commercial radio-telegraphy.

Augusto Righi was a professor of physics at the University of Bologna and a neighbor of the Marconi family in that delightful city of good pasta and sausage. Righi, a specialist in electromagnetic phenomena, redoubled his investigations when Hertz published his discovery of electromagnetic waves. He improved Hertz's oscillator that generated the waves; he immersed his spark gap—four brass balls—in Vaseline. To control the spark across the gap he placed a telegraph key in the primary circuit of his induction coil, a key which would make dots and dashes. But Righi didn't give a thought to the possibilities of wireless telegraphy, either.

Righi's work was not of world-shaking importance. What was important was Marconi's knowledge of it through his friendship with the professor. It must be pointed out, however, that Marconi never studied under Righi. Marconi never went to a university.

Guglielmo Marconi was born in Bologna on April 25, 1874, in a dismal old Italian pile called Marescalchi Palace. Giuseppi Marconi, his father, was quite a rich man. His mother was Irish, a member of the Dublin whiskey-distilling family of Jameson, and on her mother's side a Haig. It was not in Bologna, but in Leghorn, where the Marconis went to escape the rigors of the north Italian winter, that young Guglielmo first became interested in scientific matters.

In 1894 he and his older brother Luigi were lazing away the summer at Biellese in the Italian Alps. Guglielmo, bored, picked up a scientific journal which contained an article on the experiments of Hertz, who had died earlier in the year at the age of thirty-seven.

Now Marconi was bored no longer. For an idea that had somehow escaped the great electrical pundits of the day had settled itself in the head of a twenty-year-old youth: "Why not use Hertzian waves for telegraphy without wires?"

"It seemed to me," Marconi said in a lecture years later, "that if the radiation could be increased, developed, and controlled it would be possible to signal across space for considerable distances. My chief trouble was that the idea was so elementary, so simple in logic, that it seemed difficult to believe no one else had thought of putting it into practice. I argued there must be more mature scientists who had followed the same line of thought and arrived at almost similar conclusions. From the first the idea was so real to me that I did not realize that to others the theory might appear quite fantastic."

The idea that sparks from an induction coil might be used to send messages became an obsession. For the rest of the summer Guglielmo thought of nothing else and compulsively sketched wiring diagrams of what were to become the world's first wireless hook-ups.

In the fall the Marconis came home to their country seat outside Bologna, the Villa Grifone in Pontecchio. There Guglielmo organized a top-floor workshop and set to work. We don't know where Marconi got the parts for his experiments: the induction coil, the Branly coherer, even copper wire. Almost certainly not in Bologna, although he may have borrowed some components from the university. Most likely it was accumulated in bits and pieces from London or Paris.

At first Marconi had little success. His spark coil sparked spectacularly, but there was no response from the receiver. After a month or so of frenzied labor arranging and rearranging his instruments—for Marconi did not know exactly what he was about—he hit upon the right combination, and proudly showed his parents that he could ring a bell on the ground floor by pressing a switch on the top floor—and without wires.

A day or two later clicks were transmitted from one end of the house to the other; then from the house to the garden. Signora Marconi was sure her boy was a genius, but

Right: Guglielmo Marconi as a small boy, with his mother and elder brother, Alfonso. Middle: Marconi's 1896 beam transmitter and receiver. Opposite: Marconi in England in 1896, with his "black box" apparatus.

Signor Marconi was skeptical. "I'll stand near the receiver on the lawn," he said. "You tap out the Morse letter S from the house." A moment later the dit-dit-dit came through. Papa was convinced. And Papa gave Guglielmo 5,000 lire (about $1,000 in those days) to pursue his experiments.

In the spring of 1895 Guglielmo made a tremendous advance, a discovery which would make wireless practical over distances far greater than those from the villa to the garden—distances which would first encompass the earth and then reach far out into the universe and to universes beyond. It was Marconi's great basic invention—if, indeed, it was his. He built an aerial—an antenna which he connected to one side of the spark gap. (Hertz had merely used a horizontal rod ending in a plate.) The aerial was a metal cylinder atop a pole. He connected the other side of the spark gap to a ground—at first, a copper plate lying in the ground. The receiver also got an aerial and ground.

We don't know where Marconi got the idea, but several people had used such aerials before. In 1866, an American, Dr. Mahlon Loomis, sent messages a distance of fourteen miles without the use of wires. Later he signaled two miles between ships on Chesapeake Bay and got U.S. Patent No. 129,971 for "aerial telegraphy employing an 'aerial' used to radiate or to receive the pulsations caused by producing a disturbance in the electrical equilibrium of the atmosphere."

In 1885 Thomas A. Edison filed for a patent on long-distance telegraphy without wires. It wasn't issued until December, 1891. In the specifications for his patent application Edison wrote: "I have discovered that if sufficient elevation be obtained to overcome the curvature of the earth's surface and to reduce to the minimum the earth's absorption, electric telegraphing or signalling between distant points can be carried on by induction without the use of wires connecting such distant points."

Edison went on to point out that the masts of ships were ideal places for wireless aerials for intership communication. "In communicating between points on land," he wrote, "poles of great height can be used, or captive balloons. At these elevated points, whether on masts of ships, upon poles or balloons, condensing surfaces of metal or other conductors of electricity are located. Each condensing surface is connected with the earth by an electrical conducting wire."

Obviously, Edison not only had the idea, but also a patent for the aerial and the ground long before Marconi used it. Later the Marconi Wireless Telegraph Company bought Edison's patent for what was said to be "a small amount of cash and quite a little stock."

Using his aerial and ground Marconi was soon able to increase greatly the distance over which his signals could be read. He placed his receiver on a hill 1,700 meters from the villa. His brother Alfonso was stationed near the receiver with instructions to wave a flag when he saw the coherer's hammer tap out the three dots of the letter S. Marconi tapped his key thrice and the flag waved.

Next he transmitted *through* the hill. Since his brother's flag would be invisible, he gave him a shotgun. Happily, the shotgun banged its acknowledgment.

Marconi later claimed he had offered his discovery to the Italian government and had been turned down. England was the obvious place to go with it. For there Signora Marconi—nee Jameson—knew those rich and influential Victorians who could give little Guglielmo's invention a push.

Marconi arrived in London with a letter to Sir William Preece, Engineer-in-Chief of the British Post Office. Preece had been a telegraph engineer and had experimented with "wireless" signaling by induction. This involved laying huge loops of wire hundreds of feet long on each side of a body of water, each loop being grounded. A battery-powered buzzer signal in one loop induced magnetic forces which could be picked up in the opposite loop. In 1892 Preece was commissioned by the British government to experiment with wireless communication between shore stations and lighthouses. Several years later, when the four-and-a-half-mile telegraphic cable between the Isle of Mull and the mainland broke, Preece used his induction system to send messages in Morse.

Preece, an equable and fair-minded man, said he thought Marconi's system better than his own and better than that of Admiral Henry B. Jackson of the British Navy, who was conducting successful wireless tests between H.M.S. *Defiance* and the gunboat H.M.S. *Scourge* with equipment not

much different from that used by Marconi. What pressures from financiers in the City were put upon Preece to act as he did we have no way of knowing, but when asked why Marconi was declared the inventor of wireless, rather than Jackson, Lodge, or himself, Preece said: "He has not discovered any new rays. His recorder is based on the Branly coherer. Columbus did not invent the egg but he showed how to make it stand on its end."

Helped by Sir William, and with the strong backing of a cousin, Jameson Davis, the financial front man of interests in the City, Marconi got a series of tests under way. At first signals covered only a hundred yards, then between London's General Post Office and the Savings Bank Department in Queen Victoria Street. Soon Marconi was signaling across Bristol Channel, from Penarth to Weston-super-Mare, where he had set up tall masts with vertical wires as antennas.

Now a corporation, first called the Wireless Telegraph and Signal Company, and later the British Marconi Wireless Telegraph Company, was formed and capitalized for £100,000, with Jameson Davis as managing director. Patents were acquired. Soon a phalanx of famous scientists and physicists was enlisted, including Oliver Lodge and Ambrose Fleming. By 1901 there were seventeen electrical engineers working for Marconi.

Oddly, at this time Marconi's wireless men were convinced that the higher the receiving and transmitting antennas, the greater the range. To cover a yacht race for the Dublin *Daily Express,* the receiving mast at Kingston was one hundred and ten feet high. The publicity attending coverage of the yacht race resulted in as puerile an exchange of messages as may have ever been transmitted. Queen Victoria, then in residence at Osborne House on the Isle of Wight, asked Marconi to keep her in touch with Edward, the Prince of Wales, aboard the Royal Yacht at Cowes. Poor Edward had a sore knee and Her Majesty required bulletins; to this end a one-hundred-foot mast went up near the palace. The yacht's antenna went up on an 83-foot mast.

A typical message:

"from Dr. Tripp to Sir James Reid
H.R.H. the Prince of Wales has passed another excellent night, and the knee is in good condition."

But soon there were more serious uses for wireless. To test its practicality, the English Lightship Service authorized wireless communication between the South Foreland Lighthouse at Dover and the East Goodwin Sands Lightship, twelve miles away. In the spring of 1899 the Lightship Service got its proof. Heavy seas knocked the lightship on its beam ends; bulwarks were carried away. What must have been the first SOS brought help (although the actual signal "SOS" wasn't in use until years later). Two months later the steamer *R. F. Mathews* rammed the same lightship and again wireless brought lifeboats to rescue the crew.

At the same time Marconi's wireless messages crossed the English Channel. Other tests and proofs followed: wireless from shore to ship, from ship to ship. Distances became ever greater. But Marconi was ambitious. How about a wireless message across the Atlantic?

In October, 1900, construction of a station which Marconi hoped would send its dots and dashes across to North America was started at Poldhu in Cornwall. Its transatlantic receiving twin was erected at South Wellfleet on Cape Cod.

Marconi's technical ability was not equal to designing a generator which would handle the considerable power necessary to make the two-inch spark which Marconi thought he needed to span the Atlantic. The apparatus was therefore designed by Dr. Ambrose Fleming. Tuning to the correct wavelength was also necessary and Marconi's system of "syntonization" (tuning) was incorporated. This was patented under No. 7777, which later became famous for the number of law suits that were fought over it. Sir Oliver Lodge had earlier patented a system of tuning that the Marconi Company eventually found it necessary to buy.

Further, Marconi had given up his old coherer as a detector and instead used an improved type which he called "the Italian Navy Detector." Invented by Professor Thomas Tommasina of Geneva, it consisted of a tiny globule of mercury between plugs of carbon and iron in a glass tube.

The aerial system at Poldhu consisted of twenty wooden masts 200 feet high in a 200-foot circle. The masts

1

2

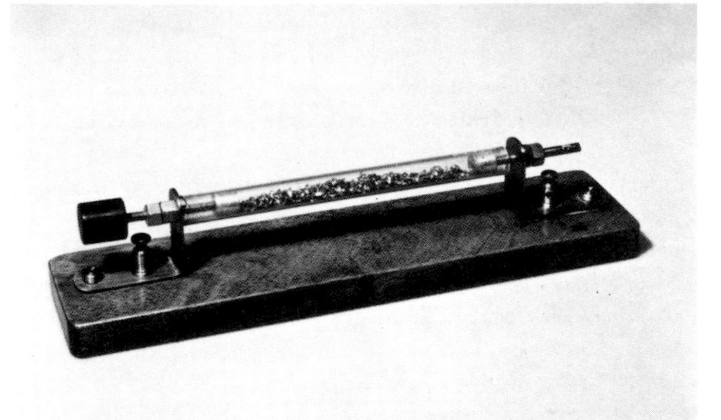
3

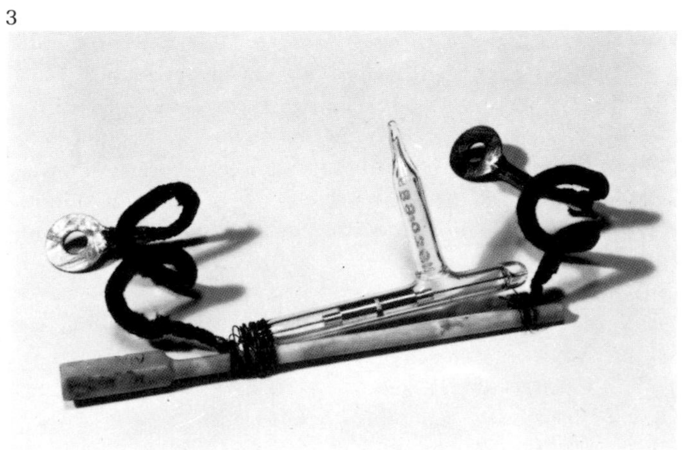
4

1. British Post Office engineers examining Marconi's apparatus during the 1897 Bristol Channel experiments.
2. Sir Oliver Lodge.
3. Branly-Lodge coherer.
4. Marconi's version of the coherer.

supported a conical arrangement of 400 wires. When almost completed, the entire giant cat's cradle was blown down by a violent storm. A few weeks later the antenna on Cape Cod was similarly destroyed. Marconi didn't wait for the complex Poldhu towers to be rebuilt. Instead, two 150-foot masts (rather than twenty) were to carry the wires between them—fifty-five of them in a fan-shaped array. In a test to Crookhaven, Ireland, 225 miles away, the raucous growl of the Poldhu spark almost blasted the ears off the operators.

Marconi then decided that his receiving station for the transatlantic test would be Newfoundland. Late in 1901, he landed at St. John's with two assistants and a load of peculiar impedimenta—captive balloons and big kites. The rather brisk weather of Newfoundland in winter made the construction of towers too chancy.

Marconi set up his station on a flat-topped hill overlooking the port of St. John's. Later, he wrote about his travails: "On Tuesday we flew a kite with 600 feet of aerial as a preliminary test, and on Wednesday we inflated one of the balloons, which made its first ascent during the morning. It was about fourteen feet in diameter and contained about 1000 cubic feet of hydrogen gas, quite sufficient to hold up the aerial, which consisted of wire weighing about ten pounds. After a short while, however, the blustery wind ripped the balloon away from the wire. The balloon sailed out over the sea. We concluded, perhaps the kites would be better, and on Thursday morning, in spite of a gusty gale we managed to fly a kite up 400 feet. . . .

"In view of the importance of all that was at stake, I had decided not to trust entirely to the usual arrangement of having the coherer signals record automatically on a paper tape through a relay and Morse instrument, but to use instead a telephone connected to a self-restoring coherer. The human ear being much more sensitive than the recorder it would be more likely to hear the signal.

"Before leaving England I had given detailed instructions for transmission of a certain signal, the Morse telegraphic S—three dots—at a fixed time each day beginning as soon as word was received that everything at St. John's was in readiness. If the invention could receive on the kite-wire in Newfoundland some of the electric waves produced, I knew the solution of the problem of transoceanic wireless telegraphy was at hand.

"I cabled Poldhu to begin sending at 3 o'clock in the afternoon English time, continuing until 6 o'clock; that is, from 11:30 to 2:30 o'clock in St. John's."

It was near noon on December 12, 1901. Marconi sat in front of his jumble of equipment, a telephone receiver held to his ear with one hand. With his other hand he fiddled with his coils, hunting for Poldhu's wavelength. At a guess Professor Fleming had put it at not less than 960 meters, but Fleming had yet to devise a wavemeter.

It was a miserably cold and foggy day, even for Newfoundland. Marconi's eyes could follow his aerial wire out the window to where it was attached to its kite straining and swaying in the murk. The noise of the waves smashing against the base of the cliff was distracting. Marconi sat sipping hot cocoa laced with Scotch and straining to hear.

The telephone clicked once. Marconi assumed it was Poldhu. He stopped tuning and stopped breathing. "Suddenly," wrote Marconi later, "at about 12:30 o'clock, unmistakably three scant little clicks in the telephone receiver, corresponding to three dots in the Morse code, sounded several times in my ear as I listened intently. . . ."

When Marconi announced his fantastic success he immediately became a world hero. The press went wild. *The New York Times* ran an editorial headed: "The Epoch-Making Marconi."

But there was at least one young American wireless engineer who was more than skeptical. He was Lee de Forest, who five years later would invent the vacuum tube, the fundamental element necessary to put an end to the crude Hertzian spark transmitter, and make radio and television broadcasting possible. When he learned of Marconi's claim to have heard those transatlantic dots, he wrote in his diary: "Signor Marconi has scored a shrewd coup. Whether or not the three dots he heard came from England or . . . from Mars, if I am aught of a prophet we will hear of no more trans-Atlantic messages for some time."

De Forest was but one of many inventors concerned

1

2

4

3

1. The first circular antenna erected at Poldhu for the transatlantic test was blown down in a gale in September, 1901.
2. The collapsed antenna.
3. Raising a kite antenna for reception tests at St. John's, Newfoundland, in December, 1901. Marconi is at far left.
4. Marconi with the apparatus used for receiving transatlantic signals in December, 1901.

Right: Marconi magnetic detector. Opposite: Wireless room of the S.S. *Lusitania* in 1907. The sinking of the Cunard liner *Lusitania* by a German submarine in 1915 helped bring the United States into World War I.

with wireless in those early years of the century. In the U.S. there was Reginald Fessenden, who invented the electrolytic detector in 1902 and who went on to invent various forms of rotary spark gaps, the heterodyne receiver, and a host of other important inventions. Ernst Alexanderson was another pioneer. Working with Fessenden he designed the alternators bearing his name which in 1906 first made speech transmission practicable, and reliable transatlantic wireless telegraphy commonplace, during the Kaiser war.

Not only did de Forest devise radio's most important invention, he was also a fascinating man. He was born on August 26, 1873, in the parsonage of the Congregational Church in Council Bluffs, Iowa. His father was the Reverend Henry S. De Forest, who was descended from French Huguenots who had settled in New England. His mother, Anna Margaret, the daughter of a Congregational minister, could trace her family back to John Alden.

In 1879, when Lee was six, the De Forest family, including Lee's sister and brother, moved to the backward, semicivilized town of Talledega, Alabama, where the tough-minded Reverend had taken on the formidable job of president of a normal school (later a college) for the black children of former slaves. The Deep South, in post-Reconstruction days, was a very rough place for the little Yankee boy. The Negroes were beyond his comprehension and both the rednecks and the upper-crust types at first couldn't understand northern whites laboring to improve the lot of blacks. The local boys gave Lee a hard time.

But he had other resources. A company of engineers arrived in the neighborhood to build a blast furnace. Lee, fascinated, built one of his own. When he was ten or eleven he became curious as to exactly how a locomotive worked. He found some of what he wanted in the *Mechanical Encyclopedia* in the college library, but the reversing gear baffled him. On a railroad yard sidetrack he came upon a deserted locomotive. He climbed under its boiler and, as he wrote in his autobiography, "observed the intricate scheme of the eccentric rods, pistons and valve rods . . . in my mind I could see the locomotive in operation. The whole method of reversing the locomotive became clear at last." Joyfully he skipped home, singing, "Oh, I am happy; I am so happy."

Inspired, he gathered paint kegs, big packing cases, tin cans, wooden strips (for drive rods), planks, barrel heads (for wheels) and, helped by his brother Charlie, spent weeks building a wooden locomotive which "became the talk of the neighborhood." And the wooden reversing gear actually moved like a real one.

Dr. De Forest wanted his son to follow the ministry, but at fifteen the boy blasted this hope. He painfully typed on his father's old Fitch typewriter:

"Dear Sir: . . . I intend to be a machinist and inventor, because I have great talents in that direction . . . if this be so, why not allow me to so study (at the Sheffield Scientific School at Yale) as to best prepare myself for that profession. . .?"

He added a few more persuasive arguments, signed his letter "your obedient son" and then added a postscript, "This machine beats Mr. Silsby's all to flinders."

At Yale, where he had an endowed scholarship, Lee de Forest (Lee's father capitalized the D; Lee and his brother used it lower case) did not do much better than so-so. He wasted time fiddling with pointless inventions. He tried to win a $50,000 prize for a new kind of subterranean trolley-car system and he spent rather too much effort chasing girls. He wanted desperately to be elected to Sigma Xi, the honorary scientific society, but didn't make it."I shall show them some day what a mistake they made," he wrote in his diary.

But de Forest was determined on a scientific career. He stayed on at Yale for a Ph.D., and chose as the subject of his research, "Reflection of Hertzian Waves from the Ends of Parallel Wires." He spent seven months in the cold, dark basement of Sloane Laboratory working with Hertzian waves, much of the time with the Branly coherer, which he called "a most erratic, undependable device."

De Forest's first job after taking his doctorate was a menial one in the Western Electric laboratories in Chicago. He was paid $8 a week. His work was in telephony, but he neglected it for experiments in wireless, especially with a new type of electrolytic detector he devised. He called it a Responder.

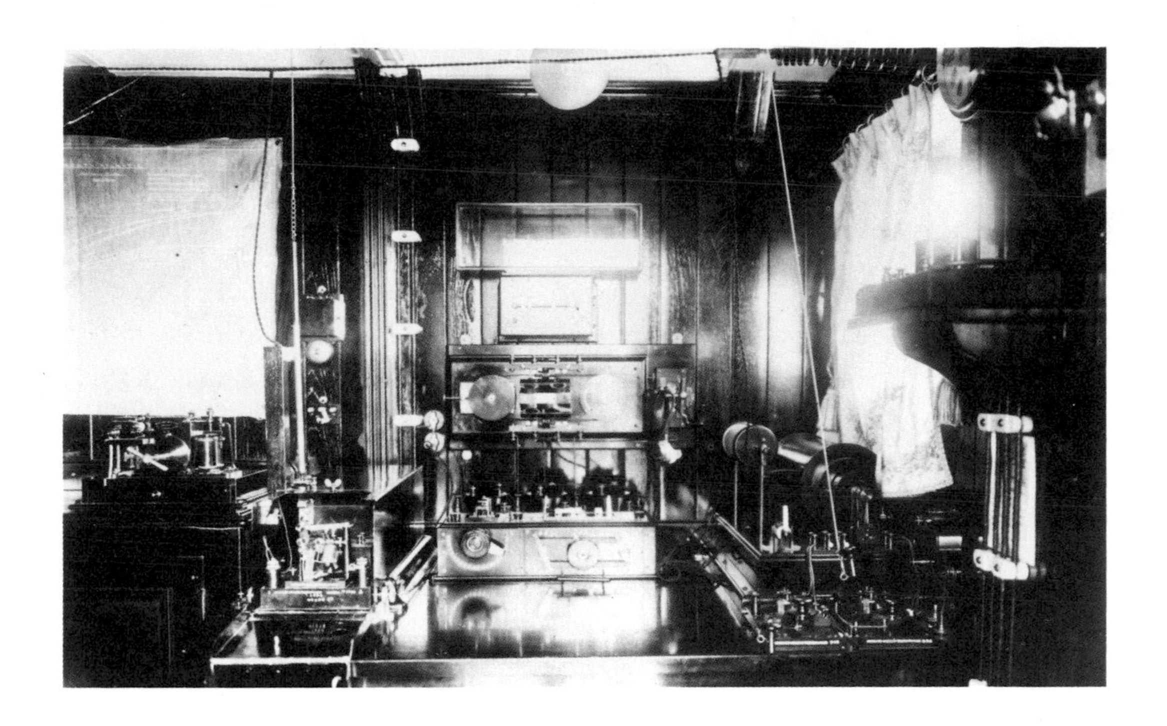

In 1900, fed up with Western Electric, he got a job with a new outfit grandly called the American Wireless Telegraph Company. Using a Branly-type coherer and an antic system dependent on puffs of air for decohering, the people at American Wireless were getting poor results. They insisted that de Forest give them his Responder. He refused and was fired.

He next went to work as an editor of a magazine, the *Western Electrician.* With two friends—brothers named Smythe—he continued to experiment after hours. It was then that he accidentally observed a phenomenon which would years later lead to his invention of the vacuum tube. He noticed that a Welsbach gas burner became brighter when the spark from a nearby induction coil leapt across its gap.

"The startling idea became firmly fixed in my mind that I had here discovered a radically new and surprising influence of electromagnetic waves upon heated gases and/or incandescent particles. . . . I would further investigate the phenomenon, and eventually apply it as a radically novel detector of Hertzian waves." Meanwhile, de Forest busied himself with improving the Responder.

In 1902 he fell in with a stock promoter named Abraham White, who set up a company called the American de Forest Wireless Telegraph Company and started selling stock—$3 million worth. He got orders from the War Department and the Navy, which preferred dealing with Americans rather than Marconi or Slaby-Arco, the big German firm which eventually became Telefunken. The United Fruit Company also had stations built in Central America.

Grandiosely, White issued high-flown prospectuses and built ninety wireless stations in Canada and the United States, many of which never sent a message. De Forest, now in New York, busied himself with perfecting a better detector. He had never heard of the "Edison effect." In 1883 Edison, noting that a molecular bombardment inside his carbon-filament light bulbs caused a black deposit on the inside of the glass bulb, tried some experiments. If he placed a coating of tin foil outside the bulb and connected a galvanometer in series with the tin foil and the positive terminal of the filament a direct current flowed through the galvanometer. When he reversed the process by connecting to the negative terminal, no current passed. When he installed two filaments inside the lamp he found that by supplying current to one, a current also seemed to be induced in the other. The "Edison effect" led Alexander Fleming to further experiments, and in 1904 he finally came up with a two-element vacuum tube, the Thermionic Valve, which had limited use as a detector but not as a relay or amplifier. Fleming received a knighthood for his invention.

It is indeed odd that de Forest, twenty years later, hadn't heard of Edison's light-bulb experiments. His experience with the Welsbach mantle led him to try using a Bunsen burner. He rigged the burner so that its flame played on two platinum electrodes, one of which was connected to the antenna, the other to the ground through a telephone receiver. This worked well enough for him to receive wireless signals from a ship in New York's Lower Bay.

"It was perfectly obvious that the gas flame would be an impractical device on shipboard, so I next sought to heat incandescent gases directly by means of electric current."

First he tried a carbon-filament lamp. No luck. Then he had a bulb made which contained both a carbon filament and a platinum plate. De Forest's plan was to connect the plate to a high-voltage source while the radio waves were to ionize the gases in the bulb and thereby make the internal resistance follow the variations of the radio signal. To increase the effect on the gas in the bulb, de Forest wrapped a piece of tin foil around the outside. This third electrode was connected to the antenna and carried the incoming signal.

"I then realized that the efficiency could be still further enhanced if this third electrode were introduced *within* [the bulb]." This bulb was the same as the previous one, with the addition of another little platinum plate to replace the tin foil that had been wrapped outside.

He theorized that the hot carbon filament was emitting electrons. These electrons bombarded the gas in the bulb and created electrically charged ions. The ion current flowed to the platinum plate which was connected to the high-voltage battery. Radio waves from the antenna impressed on the other

Right: Lee de Forest in 1907 with his first arc-type radiophone. Below: De Forest and "Honest Abe" White at the St. Louis exposition, 1904. White is the dude in the tall hat. Opposite: Thomas A. Edison holding the lamp with which he conducted the "Edison effect" experiments. He first noted the phenomenon in 1883.

5587

platinum electrode influenced the flow of the ion current so that it duplicated the incoming signal. Perhaps this control electrode, this "trigger," would work better if it were put *between* the filament and the ion-collecting plate. "Obviously, this third electrode should not be a solid plate. I supplied McCandless [the man who made bulbs for him] with a small plate of platinum, perforated by a great number of small holes. This arrangement performed much better than any preceding it but . . . to simpiify . . . the construction I decided that the interposed third electrode would be better in the form of a grid, a simple piece of wire bent back and forth, located as close to the filament as possible."

Now, in 1906, de Forest had the triode! The grid was the secret. The vacuum tube which was to make radio telephony and broadcasting practical had arrived. Nobel laureate I. I. Rabi later described it as "ranking with the greatest of all time." De Forest called it the Audion.

De Forest worked out his Audion while he was still involved with "Honest Abe" White. Meanwhile, he struggled with the multiplicity of stations White and his crooked associates had established. And he was mixed up in endless lawsuits instigated by the Marconi people.

It wasn't until 1906 that de Forest found out that his associates were looting company assets by selling them to a dummy corporation they owned. De Forest pulled out, taking with him such patents which were still pending, and including the tremendously important Audion. Later White and some of the other company directors went to jail. De Forest ended flat broke.

De Forest then formed the De Forest Radio Telephone Company, which was capitalized at $2 million. He did not use the Audion for his experiments in voice transmission since its utility in that area had not yet been discovered. Crude radio telephony at that time was accomplished by means of the Poulsen arc, which was covered by broad patents. To get around Poulsen's patents, de Forest designed a carbon-arc

Opposite: In the early 1920s the American public went wild over radio broadcasting. Millions sat glued to primitive receiving sets with headphones clamped to their ears. Right: Lee de Forest in 1921 with one of his vacuum-tube transmitters.

transmitter. Although this type of voice transmitter became a dead end when the vacuum tube was used for radio telephony, de Forest used it to broadcast music over short distances in the New York area, and in 1910 for what he claimed was the first big musical broadcast in history from the Metropolitan Opera House, including a performance by Caruso.

Actually, Ernst Alexanderson had slightly preceded de Forest, regaling ships' operators with a broadcast of a woman singing on Christmas Eve, 1906. He, of course, used one of his giant alternators to generate the carrier wave.

De Forest's Radio Telephone Company lasted until 1911, when it expired owing to his inability to raise further funds. He continued working with his Audion, however, which he tried to keep secret by sealing it inside a wooden box with the connections outside. He had not even divulged the progress he had made with it until one day while listening to an incoming message, he took his earphones—the "cans"—off and handed them to one of his operators. The man listened for a moment. Amazed by the strength of the signal he leapt from his chair and yelled, "My God, Doc, hear those signals! What have you got in that box?"

When de Forest sold his Audions to the Navy, they were still wax-sealed in boxes. The Navy operators found that they could make them work even better by feeding more than the recommended voltage to their filaments and, naturally, burned them out. Navy brass, a bit leery of the new-fangled tubes, ordered: "No more Audions; use your old detectors."

The early Audions did have a fault, a serious fault. They were gassy, "soft," and therefore far less efficient than they should have been. De Forest had assumed that the electrons flowed from the filament to the plate through a transporting medium—ionized gases remaining in the tube when it was evacuated.

But several important experimenters, including the famous Irving Langmuir of General Electric, proved him wrong. In any case, when he had first invented the Audion, vacuum pumps were not sophisticated enough to create a high vacuum.

De Forest was still forming new companies—North American Wireless, Radio Engineering Laboratories, Radio Telephone Company, etc.—and getting in and out of multifarious marriages. (Until late in life he was an enthusiastic romantic who wrote terrible love poems.) Still, he had the time and the remarkable energy for continual experimenting and inventing.

He perfected his Audion as an amplifier, and in 1913 sold rights to it as a telephonic relay to a lawyer named Meyers for $50,000. Meyers turned out to be a front for none other than the American Telephone and Telegraph Company. A.T. & T. had been prepared to pay half a million if it had to.

It wasn't until the end of 1913 that de Forest discovered that the Audion could be used for voice transmission. Now the Audion bulb—the vacuum tube—was a detector, an amplifier, and a means of transmission. But the outbreak of World War I caused all further research to be hidden by military secrecy.

Soon after the war A.T. & T., Westinghouse, and General Electric pooled their patent rights and formed the Radio Corporation of America, which then bought out the American Marconi Company. Broadcasting started from Westinghouse's experimental station, KDKA, in Pittsburgh. The rush was on.

In 1920 de Forest, who had started it all but had no stake in the new multibillion-dollar bonanza, turned away from radio and its corporate battles. In the early twenties he invented a sound-on-film system which he called Phonofilm. He couldn't sell it, and others later made billions out of a similar system.

But the boy from Talledega who built a blast furnace and a wooden locomotive kept on working by himself in a world where the lone inventor outside the regimented laboratories of the supercorporations had no chance. Still, he couldn't stop inventing. He experimented with television, high-fidelity phonographs, even short-wave therapeutic devices.

But de Forest had other pleasures, as well. He married a young movie actress when he was fifty-seven. He climbed Mount Whitney on his seventieth birthday. And he lived to be eighty-seven.

That other big man of radio, Guglielmo Marconi,

spent much of his life resting on the laurels with which he was festooned as a young man. In 1909 he won the Nobel prize for physics, and among other bushelsful of Italian medals and honors, he became a *Senatore* and a *Marchese.* He was a friend and admirer of Mussolini. He is still a hero of Italo-Americans, second only to Columbus.

He died July 20, 1937, aged sixty-three.

The Electronic Image

A few years ago millions of us around the globe were able to watch some brave fellows tramping the dusty surface of the moon, an electro-optical feat culminating the long and arduous efforts of people to "see afar"—which is what the word television means in Greco-Latin-English.

Early experimenters, who included impecunious attic scientists as well as giants like the Bell Telephone Laboratories, spent years and millions pursuing an absolutely wrong, if fascinating, course. These early inventors knew that the only way to transmit a picture by wire or radio was to dissect the image at the transmitting end into tiny bits of light and electricity, and then put the bits back together at the receiving end. Their trouble was that they did it mechanically. Today it is done electronically.

Most of the early experimenters depended on German scientist Paul Nipkow's wire-television patent of 1884, which used a scanning disc to break images into parallel lines of varying density. Roughly, scanning-disc television worked like this: In the transmitter, the object, very strongly lit, was placed in front of a rapidly revolving Nipkow disc which had a number of holes or lenses arranged in a spiral. These lenses projected a series of lines of varying intensity onto a photoelectric cell. The whole object was thus scanned by the photocell line by line. The undulations in current caused by the variations of light on the photocell were then transmitted by radio to a receiver also containing a Nipkow disc and a neon tube whose light varied in intensity with the strength of the radio signal. The neon tube's light was projected on the television screen by means of the lenses in the disc.

The writer saw a demonstration of such a primitive television system in the 1920s. He thought it was wonderful. The screen was about three inches wide. The picture was neon pink and the horizontal lines making up the image on the screen were almost a quarter-inch wide. A woman's face was just barely recognizable as such. The shadows of the eye sockets and the mouth were its only visible features. Static continually broke up the picture. Yet the very idea that moving images could be broadcast by radio and picked up at home was tremendously exciting fifty years ago.

Although many others experimented with whirling-disc TV, it was a Scotsman, John Logie Baird, who became best known for his Nipkow disc-based system of television. Baird, who had been involved with various weird and unsuccessful businesses like the manufacture of anti-sweat socks, boot polish, and soap, with a fling at jam-making in Trinidad, first started experimenting with television in 1924 in an attic in Hastings, England. Desperately short of money, Baird used an old tea chest as a base, a tin biscuit box to house his projection lamp, and other near-junk, like darning needles and lenses from old bicycle lamps. For his discs he used cardboard. His power supply consisted of old storage batteries and many small flashlight batteries. In time he got his wires and his rickety collection of bits and pieces to hold together long enough to transmit a tiny, shaky, pink image of a Maltese cross for a distance of about six feet.

Baird had enough gall to show his ridiculous apparatus and his feeble results to the press and was, through the publicity he got, able to raise a little money for further experiments. He could now afford a better laboratory and moved his gadgetry to a couple of tiny rooms in London's Soho district.

In March, 1925, department-store owner Gordon Selfridge, Jr. got wind of Baird and his experiments and managed to track him down. Baird happily gave him a dem-

Opposite: Scotsman John Logie Baird's experimental television transmitter, about 1925. Most early television experimenters used motor-driven whirling "Nipkow discs" instead of the electronic tubes used today. Baird's ventriloquist's dummy (A) was the first television performer.

onstration. Selfridge saw transmitted from one room to the next the image of a paper mask which was made to wink by covering one of the eyeholes with a bit of white paper and to open and shut the slot of its mouth by similarly simple means. Selfridge immediately offered Baird £25 a week for three weeks to show his television at the store. This was welcome not only because it gave Baird a chance to display his invention publicly—it also bought him something to eat and soles for his shoes. For Baird was literally starving.

The demonstration generated much interest. Throngs visited Selfridge's, and the prestigious British scientific magazine *Nature* devoted space to it. But the only people who were ready to help promote Baird's television were gentry who wanted him to pay them to do so. Still, the next year Baird had improved his apparatus sufficiently to invite members of the Royal Institution to his Soho rooms. The *Times* said: "The head of a ventriloquist's doll was manipulated as the image to be transmitted, though a human face was also reproduced. First on a receiver in the same room as the transmitter, and then on a portable receiver in another room, the visitors were shown recognisable reception of the movements of the dummy head and of a person speaking. The image as transmitted was faint and often blurred, but substantiated a claim that through the 'Televisor', as Mr. Baird has named his apparatus, it is possible to transmit and reproduce instantly the details of movement, and such things as the play of expression on the face."

Now Baird began to attract a little more money for his experimentation. His pictures became sharper (but still pretty awful) and he went onto a hopeless by-path using infrared "black light" to illuminate his subjects. He called this "noctovision." During 1927 Baird operated the world's first TV station—2 TV in London. It transmitted from London to Harrow—about twelve miles—on a wavelength of 200 meters. The next year he transmitted a moving image of a ventriloquist's dummy from London to New York and also to the liner *Berengaria* in mid-Atlantic. In 1929 the British Broadcasting Corporation started broadcasting Baird television programs experimentally, but discontinued them in 1935 in spite of Baird's abandonment of the disc for a "mirror drum" which gave brighter if not sharper pictures. The mirror drum used tangential mirrors, each successive mirror being orientated through a small angle so that, as the drum rotated, the area of the image was scanned in much the same way as the disc had done it.

In the United States, C. Francis Jenkins, who had been a pioneer in the development of motion pictures, Ernst Alexanderson, Herbert E. Ives of the Bell Laboratories, and others also experimented with discs, mirror drums, rotating prisms, and related kinds of whirling gimcrackery. But for all their ingenuity and the great amounts of money invested—especially by outfits like Bell, which should have known better—they were on the wrong track.

The cathode-ray tube was about to take over. It was a Russian, Boris Rosing, who in 1907 first tried to use a cathode-ray tube for television by wire. But Rosing used it only at the receiving end to paint lines of light from an electronic bombardment on the face of the tube coated with fluorescent material. His transmitter still depended on a scanning disc. Rosing's tube was based on the German Braun-Wehnelt tube of 1897, which in turn was based on the ancient Crookes tube. The Braun tube was called a "cathode-ray" tube because a beam of electrons was emitted from an electrical cathode inside the tube. It was used by electronic experimenters before the turn of the century. Rosing's method couldn't work in his day. Photocells and cathode tubes were still too crude and his weak system had no means of amplification. For Lee de Forest had not yet perfected the three-element Audion vacuum tube.

It was a remarkably foresighted Englishman, A. A. Campbell Swinton, who in 1908 suggested that television might be possible by using *two* cathode-ray tubes, one at the transmitter and one at the receiver. In 1911 Campbell Swinton fleshed out his vague suggestion in an address to the Röntgen Society. He proposed that a mosaic screen of photoelectric elements be incorporated in a special type of cathode-ray tube. The image of the scene to be transmitted was to be projected on to the mosaic screen by a lens. The back of the screen was to be scanned by magnetically controlled cathode rays. The cathode-ray beam in the receiver

1. Television picture of Felix the Cat as it appeared on the RCA-NBC version of the Nipkow-disc receiver.
2. Felix being televised.
3. Baird "Televisor." The picture appeared in the aperture at right, which contained a magnifier.
4. Engineer demonstrating a Jenkins disc-type "Radiovisor" in 1930.

was to be synchronized with that in the transmitter by means of deflecting coils.

Campbell Swinton never tried to actually build what he proposed. He, like Rosing, was ahead of his time. The tubes, amplifiers, and other sophisticated devices he would have needed hadn't yet been invented. He hadn't even heard of de Forest's Audion in 1911. Nonetheless, television today is fundamentally much like it was envisioned by Campbell Swinton.

In 1922, while Nipkow discs whirled in most television experimenters' heads, and before Baird and Jenkins and the rest had succeeded in making their scanning discs work after a fashion, an American, Philo T. Farnsworth, and a Russian who became an American, Vladimir K. Zworykin, started down the path which would lead to today's kind of electronic television.

Farnsworth, as a fifteen-year-old boy in Rigby, Idaho, had become excited by television after reading about Rosing's work in a magazine. Zworykin had been Rosing's assistant as a graduate student in St. Petersburg, and later an officer in the Czar's army during World War I. He escaped to the United States in 1919, after the Russian revolution. (Rosing was arrested during the revolution and died before he could make any further progress with his ideas.) The young American and the Russian émigré worked contemporaneously, though separately, to develop television.

It was in 1922, not long after Farnsworth read about Rosing's experiments, that he stayed after school one day to draw blackboard diagrams and amaze his physics teacher with a possible method of television using cathode-ray tubes. Four years later, at the age of twenty, Farnsworth was working as an office boy for a man named George Everson, who was running the Community Chest Drive in Salt Lake City. He happened to tell Everson about his television ideas. Everson found the electronic talk over his head, but was impressed by the young man's brilliance and decided to have people who knew more about the subject listen to him. Farnsworth then explained his concept to some electrical engineers in San Francisco who saw merit in them. Everson organized a syndicate and raised $25,000 to develop Farnsworth's television. There was a hitch when Farnsworth had to sign the legal papers. He wasn't yet twenty-one. Everson had to make a long-distance call to Philo's mother to get her permission to be appointed her son's legal guardian.

Everson rented an apartment and supplied Farnsworth with the electronic equipment he needed to work out his theories. Philo worked secretly, with the window shades pulled down. Neighbors became suspicious and the police raided the place. The Eighteenth Amendment was still in force and the cops were sure Farnsworth was running a still. They found peculiar glassware but no potable alcohol.

Crudely stated, all television systems, whether worked by scanning disc or electronically, are basically the same. The image to be transmitted is divided into tiny spots of light varying in brightness from light to dark. The spots are formed into lines rushing across the screen one above the other. The light intensity of each spot causes a change in the radio signal sent from the transmitter to the receiver, which translates the variations in radio energy received back into lines of spots to reconstitute the picture.

Farnsworth did this with a camera he called an "image dissector." The picture transformed into photoelectrons was deflected by magnetic coils back and forth and up and down past a "scanning aperture" which broke the electrons into a narrow beam which could then be amplified and sent out as a constantly changing radio signal.

In the summer of 1928 Farnsworth was able to show his backers his first results. Although only some $60,000 had been spent, he was able to demonstrate a 150-line picture scanned thirty times a second, a big step forward from the scanning-disc pictures with which some other inventors were still involved. Farnsworth's results continually became clearer and sharper.

At first Farnsworth transmitted graphic shapes and designs. A favorite of Everson's was a dollar sign which he said "jumped out at us on the screen." Later Philo experimented with sending short lengths of movie film, mostly scenes from the 1927 Dempsey-Tunney fight (Tunney went down for the memorable "long count" again and again) and the scene in the movie of *The Taming of the Shrew* in which

1

2

3

4

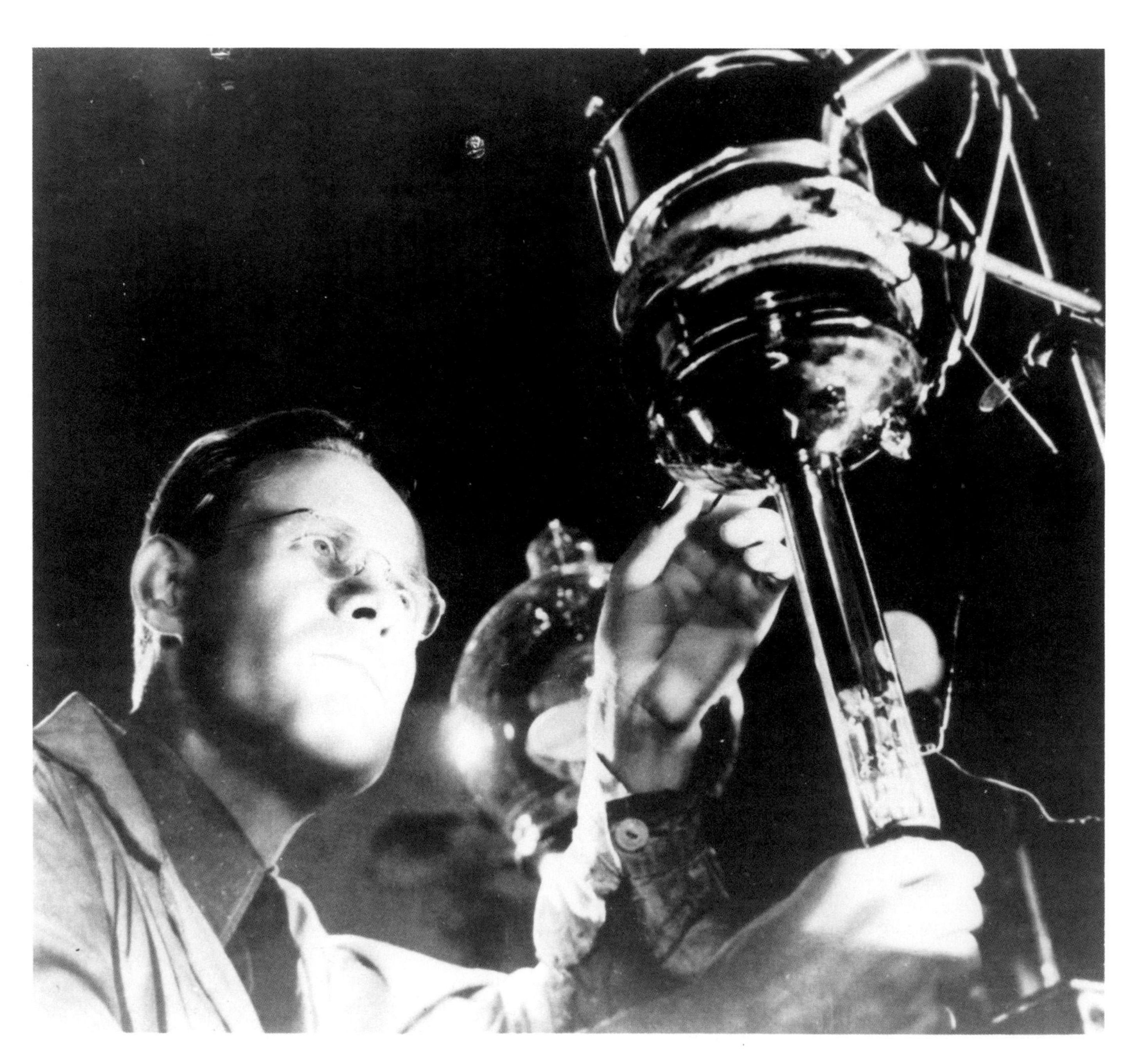

Opposite, top: Engineers assembling early TV cameras. Bottom: Philo T. Farnsworth working on a new type of TV tube in 1957. Left: Vladimir K. Zworykin with his "iconoscope" tube.

Mary Pickford combed her hair. The hair was combed endlessly for Philo's "Telecine"—his apparatus for scanning movie film.

When Farnsworth applied for an electronic television patent he really shook RCA, whose laboratories under Vladimir Zworykin had long been struggling with the problem. RCA challenged the application. Its lawyers, in proceedings claiming interference, sharply questioned Farnsworth for many hours, but failed to break him down. He got his patent in 1930, when he was twenty-four years old. A few months later Zworykin came out to California to see what Philo was up to in his laboratory. Later Zworykin was said to have claimed that RCA wouldn't need anything Farnsworth had done. Then RCA's tough chief, David Sarnoff, came to take a look. He echoed Zworykin. But later RCA found that it very badly needed some of Farnsworth's patents and paid for rights on a royalty arrangement.

During the early 1930s Philco became Farnsworth's chief backer. Although Farnsworth was able to show a remarkably clear image of over a foot square to the Franklin Institute, Philco became restive as expenses mounted. When costs passed a quarter of a million, a lot of money in Depression days, Philco pulled out.

Farnsworth's money men tried to sell his patents outright in 1938. But the price, due to high development costs, was over a million dollars and they were unsuccessful. The syndicate then decided to go it alone and bought the Capehart Corporation in Fort Wayne, Indiana, to manufacture television sets. But Pearl Harbor ended any such possibilities and Farnsworth's company built radar equipment instead.

Zworykin, of course, had been interested in television since his days with Rosing in St. Petersburg. Once in America he soon got a job doing radio research in the Westinghouse laboratory in Pittsburgh. (Westinghouse and General Electric controlled RCA until 1932.) He tried talking his bosses into letting him work out his electronic theories, but Westinghouse couldn't have cared less and Zworykin quit.

Early in 1923, after a new manager, Samuel Kintner, took over Westinghouse research, Zworykin came back. For Kintner was more receptive to his ideas. Zworykin immediately got cracking and that same year applied for a patent on his version of electronic television. By 1931 he and his considerable staff had perfected a system which was rather different from Farnsworth's method. He called his camera tube an "Iconoscope."

Today, almost half a century later, Zworykin's system, much perfected by squadrons of electronic engineers, is still basically the same. Briefly, it works like this: The image being televised is focused by a camera lens on an evacuated tube in the television camera inside which is a small screen covered with approximately 367,000 microscopic photoelectric elements forming a "mosaic." The varying light from each part of the image being televised falls upon these "dots" and gives them an electrical charge, the strength of which depends upon the amount of light falling upon the individual dots. Thus each dot becomes a tiny storage battery and the image is formed in a pattern of electrical charges on the mosaic.

The mosaic is scanned by a tiny beam of electrons no larger than a pinhead, moving from left to right and progressing downward, in the same way you read a page. This process is repeated sixty times each second, and the horizontal lines of alternate scanning are interlaced so that thirty pictures of 525 horizontal lines are produced each second.

As the electron beam strikes each dot in the mosaic, the dot is discharged through the electron beam, and the electrical impulses produced are used to modulate the signals of the TV transmitter. Each time the dots are discharged by the electron beam they are recharged by the light produced by the succeeding images falling upon them. The succession of individual "still" images creates the illusion of motion in the same way that succeeding frames on movie film do.

The receiving systems of both Farnsworth and Zworykin and those of less well-known inventors were basically not too dissimilar. All showed their pictures on the face of a cathode-ray tube—the picture tube.

The incoming succession of radio signals, after amplification, is impressed on the picture-tube grid. The pic-

1

2

3

1. TV broadcast from the 1938 auto show.
2. RCA-NBC television cameramen at the 1939 New York World's Fair.
3. President Franklin D. Roosevelt on a television screen during the broadcast at the opening of the 1939 World's Fair.
4. The picture tube stood vertically in early RCA sets. Viewers saw the pictures in a mirror.

ture tube also contains an electronic "gun" which shoots out a tiny beam of electrons which moves from left to right and progresses downward on the face of the picture tube.

The face of the tube is coated with a fluorescent material which gives off light at the spot where it is hit by the electron beam. When a TV signal is received by the grid of the picture tube, the grid controls the strength of the electron beam and thus the amount of light on the face of the tube. If the scanning of the electron beam in the picture tube is synchronized with the scanning of the electron beam in the TV camera, the picture tube will reproduce the lights and shadows of the televised image and the rapid succession of such images will give the illusion of motion.

The picture you see is actually produced by a flickering spot of light moving rapidly across and down the face of the picture tube. You see the whole picture because the screen glows for a tiny fraction of a second after the electron beam has passed. The phenomenon of "persistence of vision" makes your brain think that the picture is there all the time.

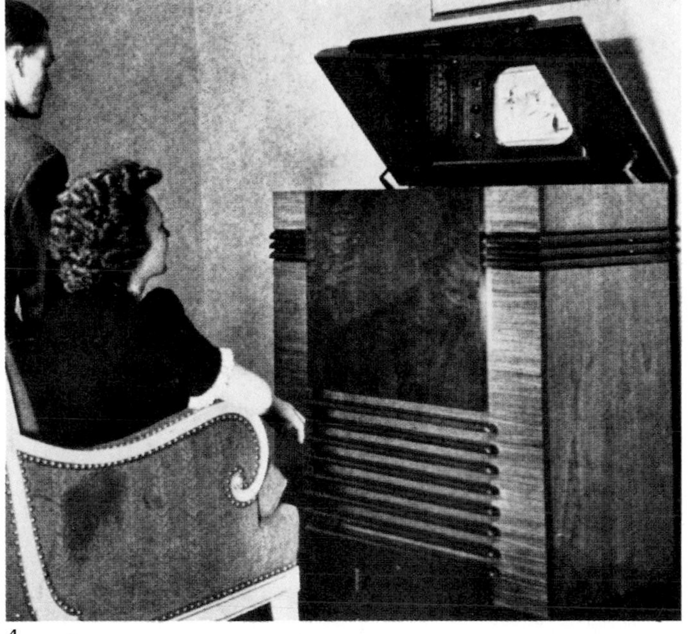
4

By 1932 NBC, RCA's radio network, was doing experimental telecasts from the Empire State Building tower. Three years later David Sarnoff dramatically trumpeted that he was about to put a million dollars into television program demonstrations.

Television for everybody had its debut at the 1939 World's Fair in New York. In February there had been an experimental warm-up telecast from the not-yet-finished fair grounds. The stars of this show were Amos 'n' Andy in blackface. On opening day, April 30, Franklin D. Roosevelt appeared on the few sets spread around to friends of Sarnoff and RCA.

RCA television sets were offered for sale soon afterward, with five-inch and nine-inch tubes. Later there were twelve-inch tube sets with the tube set vertically and viewed in a forty-five degree hinged mirror. Prices ranged from $199.50 to $600.

People had been hearing about television for decades but when it finally came they stared at the picture tubes in shocked wonder. In 1939 they could see few programs: one a day from RCA and intermittently from CBS and Dumont. By May, 1940, twenty-three TV stations were on the air in the United States. By the time war broke out in 1941, ten thousand sets had been sold, but production soon stopped. Electronic experts and production facilities went into that hush-hush device, radar. By the summer of 1946 you could again buy a television set and the boom was on. TV stations grew in number as fast as the wild forests of antennas on roof tops.

Now, thirty years later, television encompasses our earth. For instant views from distant places it is bounced off man-made satellites. Space probes take television pictures of our sister planets. In Mongol yurts and Eskimo igloos people sit watching the pictures on the fluorescent screens of their cathode-ray tubes.

But after all the mighty labors of Nipkow, Rosing, Campbell Swinton, Baird, Jenkins, Farnsworth, Zworykin, and the rest, is the stuff worth watching?

Mend or Stand at LIVERY
A FINE STUD of REAL
HORSES To BE SOLD as cheap
as dogs meat — The present
proprietor going into the Hobby
LINE
He seems to me Jack not to have quite Barrel enough
The Hobby Horse

I'll warrant him sound Sir & free from Vice
I can see he has been down once or twice though my lad
by Jupiter
4
Pubd. July 25th 1819 by G Humphrey 27 St James's St.

4 Putting the World on Wheels

Without the Horse

The automobile created a world revolution. It changed people's lives more spectacularly than any invention since man first learned to apply mechanical power to lighten his labors.

The motor car not only made it possible to travel the terrrestrial globe at will, it also rearranged the world's economic and social organization. The manufacture of cars and of the petroleum-based fuel they required became gigantic industries and spawned a myriad of subsidiary suppliers of steel, rubber, glass, textile, chemical, and electrical components. The promotion and sale of cars, the financing and insuring of cars, the maintenance and repair of cars, and the resale and scrapping of cars, not to mention the engineering and construction of highways and the care, feeding, and entertainment of travelers, created thousands of jobs that had never existed before; in fact, in all its ramifications, the automobile industry became the principal employer of every nation it invaded and conquered.

In terms of social mobility, its effect was explosive. The railway had begun the process of liberating the lifelong dwellers of country towns and villages from their bondage—while, incidentally, also giving people a taste for travel at better than sixty miles per hour. But it was the automobile that killed provincialism forever. Everyman's driveway was connected to an international network of roads. The resident of Maine could—eventually—drive to the Panama Canal, the English visitor to the toe of Italy, the Munich burgher to Norway's northernmost cape to take color photographs by the light of the midnight sun. Continents have become neighborhoods, and only the remotest corners of the world do not have an automotive link with the rest of civilization.

The automobile was probably the greatest single force in the creation of the suburb. The bicycle first enticed the city resident to venture beyond the edges of metropolis, and railroad routes bound cities and towns together like beads on a string. But the bike was dependent on muscle power, so that its comfort was minimal and its range inevitably limited. The railroad was fine for long-distance travel and of incidental benefit to the commuter. The automobile, however, could be driven for pleasure, wherever one wished, and once this new sense of freedom was fully savored, motorists were insatiable in demanding more and better highways. Most of all, the car enabled people to drive to work. It was no longer necessary to live close to the job, and millions of car owners departed the city and settled in suburbs.

Inventors had experimented with various self-propelled, road-going vehicles for nearly a century before advances in technology and the contributions of other inventions finally made the automobile a practicality in the late nineteenth century. And if the motor car was ready for use, people were ready to use it. Complex mechanisms were becoming commonplace and were no longer baffling or surprising to the citizenry.

At first the motor car was the toy of the rich. Then the middle classes—small businessmen, successful farmers, doctors ready to abandon the horse and buggy for making house calls—began to buy cars. By the outbreak of World War I mass production, mostly of the Model T Ford, had made cars readily available to Americans. A decade later the ordinary man in Britain and Europe became a buyer, too.

Automobiles soon became the proudest possessions of millions of people around the world. Their style, performance, model changes, textures, colors, and accessories are preoccupations and symbols of social status in virtually every country and on every financial level. People continually discuss cars, read about cars, spend countless hours fussing with cars, and a good part of their lives driving cars.

At the same time, the marvelous machine has its detractors. They point out, quite rightly, that it burns too much of the planet's nonreplenishable petroleum, contaminates the atmosphere, and is responsible for burying an ever-increasing area of the earth's surface under highways and parking lots. They also question whether the inefficient

Preceding pages: An 1819 English cartoon quips that the horse dealer will become the *hobby*-horse dealer. Opposite: Lady on a "Pilentum," a treadle-operated tricycle. Left: Macmillan's pedal-operated rear-driven bicycle of 1839.

internal-combustion engine should continue to power automobiles as it has for nearly a century.

What impact these protests will have is as yet unclear. As a matter of expense alone, people seem to be favoring smaller cars and more efficient hydrocarbon engines. But we have no way of divining the shape of cars to come. They may be powered by nuclear energy, steam, electricity, or the sun. Only one thing is certain. Generations still unborn will enjoy driving them.

On Two Wheels Plus Muscles

When the first white men to be seen riding bicycles pedaled into a central African village, the black natives took a most superior attitude. "Look at those lazy whites," they jeered, "they have to sit down when they walk." Today the grandsons of those Africans not only sit down while they walk, they don't even move their legs while astride their Hondas and Kawasakis.

The bicycle, one of the more ingenious and remarkable inventions of all time, was no one's idea. Nobody, in the beginning, had the daring thought that a person might balance himself on two wheels, while applying leg power via pedals and a chain to one of the wheels in order to propel himself forward. Such a wild proposal would obviously be against all the laws of God and gravity.

Yet the bicycle, which permits its rider to move at a speed five to ten times as rapid as a walking or running man, and with no greater exertion, is today the most common form of transport on earth.

Although what look like bicycles appear in bas reliefs of ancient Babylon, Egypt, and Pompeii, and a drawing of what looks like a chain-driven "safety" bicycle has recently been found in Leonardo da Vinci's fifteenth-century *Codex Atlanticus,* the modern bicycle stems from a clumsy device astride which the Comte de Sivrac propelled himself, to the amazement of strollers, in the gardens of the Palais Royal in Paris in 1791. His mount was a small wooden horse set on two wheels which he moved forward by thrusting at the ground with his feet. He could move only in a straight line since the front wheel was unsteerable. We don't know if the brave count invented this *célerifère,* but he at least had built a mansize model derived from a similar child's toy. The *célerifère* achieved a certain popularity during the Directorate and got a new name, the *vélocifère,* when the *Incroyables*—the fops and dandies of the period—took up riding the clumsy two-wheelers as a fashionable sport. By 1804 the *vélocipèdes,* as the riders (*not* the machines) were called, were holding races on the Champs-Élysées. But the agile young blades seem soon to have abandoned wheeled hobby-horses for real horses in Napoleon's cavalry. For the *vélocifère* doesn't surface again until after Waterloo.

In 1816 that prolific inventor, Joseph Nicéphore Niepce, who had earlier been involved with steamboats and engines, and who later took the world's first photographs, showed up in Paris's Luxembourg gardens with a much-improved *célerifère* which no longer looked like a hobby-horse, though it had a carved animal head up front and much bigger wheels. Niepce's first machine was faster and easier to ride than the eighteenth-century models, but was still unsteerable.

The first more nearly practical two-wheeler was built in 1817 by a German: Baron von Drais de Sauerbrun. With a steerable front wheel, a soft saddle, and a padded rest for the rider to lean against as he stabbed at the ground with his feet, it was capable of surprising speed. Von Drais, who was involved in forestry, had often to make longish trips along woodland roads and once propelled himself from Karlsruhe to Schwetzingen, a normal three-hour hike, in an hour. Admittedly, it was mostly downhill. After von Drais showed his *Draisienne* in Paris, a strong-legged Frenchman went from Beaune to Dijon, a distance of thirty-seven kilometers, in two and a half hours, averaging fifteen kilometers per hour.

The *Draisienne* immediately became *the* thing among the youthful beau monde of Paris and the Regency bucks of London, who named it the dandy-horse or hobby-

In 1819, when the *Draisienne*, or hobby-horse, became a fad of the leisure class, English cartoonists rushed to their drawing boards.

horse. Americans even tried riding them over the execrable cobbled streets of New York and Philadelphia, though without much success. Caricaturists such as Cruikshank and Rowlandson found them wonderful subjects for ridicule, and John Keats called the dandy-horse "the nothing of the day."

The public took a jaundiced view of these toys of the sons of the rich, among other things because of their arrogant and annoying habit of riding on sidewalks. Blacksmiths, fearing a decline in horseshoeing if wheeled "horses" became popular, sometimes attacked hobby-horse riders, pushing them off their machines which they then smashed with hammers. And some of the early cyclists abandoned their mounts, blaming the hernias some of them contracted upon the unnatural position a rider had to assume, the weight of the machine a hobby-horser sometimes had to lift, and the bumping he had to absorb from the iron-tired wheels.

Although the hobby-horse craze died before 1830 and cycling didn't really revive until the 1860s, there were a few efforts toward building bicycles which did not depend on riders thrusting their feet against the ground. In 1839 Kirkpatric Macmillan, a Scotch blacksmith, built a machine with foot-operated treadles which worked long rods attached to a crank at the hub of the rear wheel. This was the first real bicycle which could be balanced while the rider pumped the treadles. Macmillan proved its practicality by taking quite long trips aboard it (and being fined five shillings for knocking down a child at the end of a forty-mile trip from his home town of Courthill to Glasgow), but it had no real success.

The first popular treadle-driven machines were tricycles and quadricycles. These again were called velocipedes, and first appeared in the late 1840s. But these, too, were crude, heavy, and of limited usefulness. Still, there were some enthusiasts willing to try out the new method of wheeled progression. In 1849 the great Thomas Carlyle talked an Italian portrait painter into joining him aboard a treadle-operated tandem quadricycle for a twelve-mile trip from Chelsea, in London, to Wimbledon and back. Carlyle's wife, in a letter, wrote: "Three hours that strange pair were toiling along the highways." They must indeed have labored mightily if their quadricycle was like the one built in 1851 by J. Ward of London for the Prince Consort. It had an iron frame and tiller steering, and weighed a hefty 115 pounds.

It is hard to imagine that among the thousands of mechanically oriented men of the mid-nineteenth century, nobody seems to have had the obvious and simple idea of attaching a pair of pedals to the front wheel of a bicycle.

It was Pierre Michaux, a Parisian builder of baby carriages, who first constructed such machines for *vélocipèdistes.* According to Henry Michaux, one of Pierre's sons, a customer brought a hobby-horse to the shop for repair. Another son, Ernest, took the primitive cycle out for a test ride. When he came back Michaux *père* got the bright idea (a workman in the shop, Pierre Lallimont, later claimed he thought of it first) that a cranked axle "like the crank-handle of a grindstone" might be fitted to the front wheel so that it could be revolved by the rider's feet. Ernest did so and a French national monument at Bar-le-Duc commemorates father and son as *"inventeurs et propagateurs du vélocipède à pédale"* in 1861.

The Michaux firm developed the new pedal-operated velocipede, which later was called the "bone-shaker" in England and America. By 1865 it was turning out some four hundred machines a year.

The Michaux bone-shaker was hardly the kind of vehicle a modern cyclist would care to mount. Unlike later bicycles, its front wheel was only a little taller than the one aft. Both were roughly three feet in diameter. And they were very hard-riding wheels with wide, flat, iron tires. The frame was made of wrought iron. Only the long leaf spring on which the saddle was mounted helped ease the jolts to the rider's vertebrae. A shoe-type brake on the rear wheel was applied by rotating the handlebars. Nor was the bone-shaker light. A typical one weighed some sixty pounds.

In spite of its crudity and its weight, which exhausted its devotees, the bone-shaker remained popular until the early 1870s. Races were held, cycle clubs were formed, and magazines dealing with the sport appeared, among them *The Velocipedist* in New York. Toward the end of the 1860s bone-shakers with tubular frames, solid rubber tires, and even primitive gearshifts appeared.

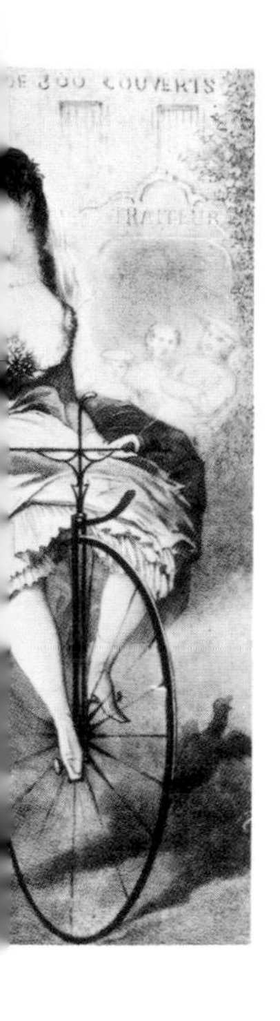

Opposite: The "bone-shaker," or velocipede, was popular in the United States in 1869. Middle: Gay blades and girls have a high old time aboard velocipedes in the 1860s. Left: A special model of the "penny-farthing" high-wheeler enabled women to ride sidesaddle and thus preserve their modesty.

And in 1868, a portent of a noisy future, a Michaux velocipede became the world's first motorcycle when a small one-cylinder Perreaux steam engine and boiler was attached to it.

France had been the leading cycling country, but in 1870 its interests were diverted to the Franco-Prussian war, and cycling declined. Others, especially the British, continued the evolution of the bicycle.

It was in Coventry that the bicycle received its next impetus. Over the years the Michaux bone-shaker's front wheel had grown a bit larger than its rear wheel, but in 1872 the British bicycle's front wheel suddenly grew to an amazing diameter while its rear wheel shrank to the size of a caster.

This kind of bicycle, at first known as the "ordinary" (though it looked most extraordinary) and later called the "penny-farthing"—the different sizes of the wheels being related to those of the coins—achieved a most undeserved popularity. For the penny-farthing was a dangerous brute of a machine which only athletic young men could mount. The front wheel was usually about 54 inches in diameter. And normal young blades envied those tall fellows who had legs long enough to reach the pedals of 60-inch wheels. For wheel diameter was a measure of esteem among cyclists. This meant that the seat located atop the front wheel was some five feet off the ground.

The ordinary was as unstable as it looked. True, its big front wheel smoothed small holes and bumps in the road more than the small-diameter wheel of a modern bike, but a stone or a tree branch in the road could cause instant disaster. Too-sudden braking (by means of a spoon-shaped device applied to the top of the front tire) brought the front wheel to a stop, but also caused the rest of the machine, including the rider, to rotate forward around the front axle in a spectacular and often painful somersault.

The Earl of Albemarle in the 1866 Badminton Library of Sport volume on cycling wrote that "the peculiar form of tumble that ensues is known by the distinctive name of 'the cropper' or 'Imperial crowner.' " Which meant that the rider usually landed on his head.

The hazards implicit in the penny-farthing did nothing to deter its enthusiastic riders. The high-wheelers continued to flourish and to attract the attention of inventors bent on improving them. Ball bearings, tangentially spoked wire wheels (invented by James Starley in Coventry), hollow steel tubing for frames, sprung seats, and hollow rubber tires (not yet pneumatic) made the penny-farthing lighter and more comfortable, if no less lethal.

In 1876 Colonel Albert Pope, the same American who later formed the Pope automobile empire, started to import high-wheelers into Massachusetts and soon afterward began building them. His Columbia bicycle is still a big name.

The high-wheeler became a sensation among Americans, but its cost kept it from becoming a boon to the walking poor. The penny-farthing's $300 price would be about $2,000 in 1970s currency. A workman in those days rarely made $10 a week.

The high-wheeler became the beau ideal of young sports in Britain, France, and the U.S. Clubs and racing proliferated, and the venturesome attempted seemingly impossible tours—from the Atlantic to the Pacific, from Land's End to John O'Groats. And super-toffs like the Prince of Wales went to bicycle races.

In an effort to prevent the dreaded "header," an American outfit, the Smith Machine Company, hit upon an odd stratagem. It turned the high-wheeler back to front, putting the small wheel in front—the farthing ahead of the penny. It did prevent head-busting somersaults, but it made for rather tricky steering.

But the ordinary's twenty years of popularity ended when the "safety" bicycle appeared in 1884.

Even before that there was a flurry of interest in tricycles. The tricycle attracted those less-agile citizens who feared mounting and balancing the tall high-wheelers. Further, there was no need to dismount when coming to a stop. But tricycles were not really quite so safe as they seemed. They too had wheels of enormous diameter, and in case of a spill it was difficult to get clear of the spokes. Nor was it difficult to capsize a tricycle if a corner was taken a mite too fast, or if the rider swerved to avoid a rut or stone while charging downhill.

Right: U.S. letter carrier pedals a high-wheeler on his appointed rounds, 1888. Opposite: Gentleman takes to the road aboard his Starley "Royal Salvo" tricycle in 1885. Tricycles required no balancing, but were heavy and poor hill-climbers.

Tricycles in the 1870s and 1880s were of many configurations. The first commercially produced model—the Dublin of 1876—had a large, 38-inch driving wheel aft and two front steering wheels. The driving wheel was turned by a system of wooden treadles, levers, rods, and cranks. The operator sat high above the front wheels on a seat mounted on coil springs. James Starley's 1878 Coventry Lever tricycle had its 50-inch driving wheel at one side amidships. The smaller, steerable wheels were on the other side—one in front and one behind the big driving wheel. Other manufacturers offered other combinations: steering wheels aft, forward, 60-inch driving wheels, pedal drive, chain drive, tiller steering, handlebar steering, lever steering. And there were dozens of ways and places to accommodate riders. For tricycles often carried twosomes, either in tandem or side-by-side (these were called "sociables"). And an extra rider often was welcomed to do part of the hard work of pedaling, for tricycles were heavy. A light "lady's" tricycle of 1880 might well have weighed more than a hundred pounds.

In the eighties there was a flurry of even more antic machines: dicycles, in which the rider sat between two huge wheels. One, the Otto, was made by the Birmingham Small Arms Company, which produced about a thousand of them. Pedals and belts drove the wheels. Steering was effected by slackening one or the other of the belts, so that one wheel went faster than the other. But you had to learn to balance the rig in a fore-and-aft direction.

Quadricycles—four-wheelers even heavier than tricycles—had a short vogue among people strong enough to pedal them.

It was on a tricycle that the first pneumatic tires were fitted. In 1888 Dr. John Boyd Dunlop of Belfast, Ireland, presented his boy Johnny with a tricycle. Johnny complained that its hard tires made the trike uncomfortable and difficult to pedal. Dr. Dunlop had the odd hobby of making himself gloves of rubber and canvas, and got the idea that similar "gloves" filled with air might prove more resilient than the solid rubber tires on the trike's wheels. Using strips of linen from one of his wife's old dresses and bits of rubber sheeting, he contrived a set of pneumatic tires. A cycling friend, William Hume, talked Dunlop into making a couple of similar tires for his racing bicycle upon which he then proceeded to clean up in racing. Within a very few years the solid tire was dead—and Dunlop went on to become a famous name in automobile tires the world over.

It was the Rover "safety" bicycle which J. K. Starley invented in 1885 that changed the world of cycling. It no longer had a front wheel of huge diameter and it had pedals amidships which drove its rear wheel by way of a chain and sprockets. It was safer to mount and easier to balance. The French claimed that it had been originally conceived by one André Guilmet in 1869, but was neglected because Guilmet was killed during the Franco-Prussian war. When found in a Paris loft in about 1880, the Guilmet model had wire wheels, rubber tires, and a chain drive. Either these elements were added after Guilmet's death, or he had been, in fact, far ahead of his time.

In 1879 Englishman H. J. Lawson introduced a machine he called a "bicyclette" (a name the French still use). Although the bicyclette's front wheel was somewhat larger than the one in the rear (40 inches to 24), the disparity was not like that of a high-wheeler. But the Lawson bicycle must have been slightly ahead of its time; perhaps it didn't look right to people. Anyhow, when Rudge put it on the market it sold poorly. It wasn't until 1885, after several false starts, that Starley's Rover caught on. It was so modern in concept that few people would pay any attention if one were pedaled down the street today. True, it had bigger wheels than those on a modern bike. The front wheel was 32 inches in diameter, the rear wheel 30 inches. Further, its top tube and seat tube were, to our eyes, oddly curved. And the 1885 Rover safety was no lightweight. It weighed a hefty thirty-seven pounds and it still rode on solid rubber tires.

By 1890 the tubes were straight. The "diamond" frame, introduced by the English Humber company in its model of that year, is still almost universally used. The Humber also had a raked, ball-bearing steering head, a spring saddle on an adjustable tube, and wheels of similar diameter. The modern bicycle had arrived. Gearshifts, various forms of brake, and the rest of our modern appurtenances and im-

Right: 1878 Coventry Lever tricycle, built by James Starley. Middle: The handlebars of the 1880 Starley Rover were almost amidships, and they steered the front wheel via rods. Opposite: The 1885 Rover, an early "safety" bicycle, looked almost like a modern machine. Note, however, that the front wheel is still larger than the rear.

provements were but delicious frostings on the cake.

The advent of the safety bicycle caused a veritable explosion of enthusiasm for cycling. By the mid-1890s there were some four hundred bicycle builders in the United States alone. Cyclists by the hundred thousands formed clubs to protect their interests, to organize tours, and above all to lobby—successfully, as it turned out—for better highways. By 1898 the League of American Wheelmen had a membership of 102,636, and almost every fair-sized town had a local chapter.

The bicycle not only set the stage for the automobile by imbuing hundreds of thousands of cyclists with the desire for farther-ranging travel without muscle-wearying pedaling, but its builders also provided the workshops which became the nesting places for many future makes of motor car. In England Hillman, Singer, Humber, Rover, Lea-Francis, among others; in Belgium Minerva; in France Peugeot; and in America Winton, Pierce, Pope, and Thomas come to mind as a few makes that were bicycles before they were cars.

And there was once a Wright Brothers bicycle.

The bicycle retained its importance in many countries, especially in Scandinavia and Holland, where it became perhaps the most important type of private transport. (In Denmark there is still one bicycle for every two people.) And although bicycles by the multimillion became common in the Far East, cycling waned somewhat in Britain and almost died in America. In the U.S., of course, the automobile became king and only children rode bicycles—very bad bicycles which had degenerated into heavy, fat-tired toys.

In Europe, largely due to a continued enthusiasm for racing (the Tour de France, for example, is still France's most important sporting event), the bicycle was continually improved. The use of English Reynold's tubing and similar, if less famous, structural material made it possible to build ever lighter and stronger frames. Gearshifts, like the British Sturmey-Archer three-speed hub gear, first patented as early as 1902, and later the French *derailleur* gearshifts (which go back to 1909) in which the chain is moved from gear to gear, took much of the pain out of hill climbing. Lighter, stronger wheel rims, better tires, new forms of caliper brakes contributed to the transformation of the bicycle from the heavy forty-pound muscle-wearying safety of the turn of the century to the twenty-pound delight it became by, say, 1965.

I choose that date because it was then that Americans seem suddenly to have become aware of the modern ten- or fifteen-speed, well-braked bicycle. Until then the few adult Americans who cycled had been using the heavy three-speed "English racer" which wasn't a "racer" at all, albeit far superior to the usual crude child-oriented ironmongery product turned out by American factories. Now ten-speed light bicycles are sold by the million—some thirteen million in 1973—and bicycle paths, sew-up tires, *derailleurs,* and cotterless cranks are hot subjects for discussion at cocktail parties and in high-school lunchrooms from Portland, Oregon, to Portland, Maine.

Four Wheels and an Engine

The bicycle was one of the parents of the modern automobile. The other was the internal combustion engine. Yet the earliest self-propelled vehicles to run on roads had neither of these ancestors. They were relatives of the steam locomotive and their progeny were the steam automobiles whose line died out early in this century.

Nicholas Joseph Cugnot, a Swiss working for the French army, built the first vehicles to move under their own power. These machines were hardly means of personal transport, for their purpose was the hauling of artillery. Cugnot started designing his *fardier* in about 1765 and had the same problems that Newcomen and Watt in England had struggled with—eighteenth-century technicians' inability to cast cylinders and pistons accurately, and the impossibility of getting anyone to bore a truly round hole. Still, by 1769, Cugnot's steam tractor, built by a Monsieur Brezin, was able to steam under its own power, attaining a speed of about three miles an hour with four passengers aboard. But its boiler was incapa-

ble of supplying enough steam and after about fifteen minutes of running a stop had to be made to build up more pressure to the quite high level of sixty pounds per square inch— remarkable in those days of the low-pressure engines.

Cugnot seems also to have had steering problems. The stories about him roaring through the streets of Paris, terrifying the fleeing populace with his fire-breathing monster until he smashed it into a wall, are the romantic inventions of nineteenth-century writers. Nor was Cugnot ever locked into a jail cell for his imaginary ride. It does seem true, however, that a wall was knocked down. Although this may prove that Cugnot's creation was a bit difficult to steer, it also shows that it wasn't entirely lacking in pushing power.

For some reason a second version of Cugnot's steamer was built at the Royal Arsenal to the order of the king's minister of war in 1771. This second Cugnot car now sits splendidly in a cathedral-like hall in the Conservatoire des Arts et Métiers in Paris. It's an impressively big machine, almost the size of the tractor part of a modern tractor-trailer truck, although it has only three wheels. All of the machinery is hung on the single front driving wheel, which also does the steering. To steer meant not only turning the thick and heavy wooden front wheel, but also the twin thirteen-inch cylinders (of fifty-liter capacity), the big copper boiler filled with water, and also the heavy mechanism—a ratchet and pawl system—by which the pistons' connecting rods transmitted power to the wheel.

The chassis is as crude and heavy as the framing of a barn and built of similar hand-hewn timbers. Mounted on these beams is a bench for a driver, in front of which is a steering column, atop which is a cross bar with a vertical handle on each end. Beneath the seat is a basket for fuel—coke, perhaps. Iron ladders help the driver and passengers—probably the gun crew—aboard. It's doubtful if this second Cugnot tractor ever ran. The steam pipes seem never to have been hooked up and essential parts are missing, not because of attrition but because they were never made.

As we've seen earlier, steam was the great new force remaking the world in the new nineteenth century—the century of machine power. In every advanced country, imaginative engineers were attacking the problems of constructing efficient steam-propelled road vehicles: Bozek in Prague, Evans in Philadelphia, Pagani in Bologna, Dietz in Paris, Dallery in Amiens, and others.

But it was in Britain that the greatest successes were achieved. For England was the home of steam power, the place where Newcomen and Watt had put it to work. And above all it was where that Cornish genius, Richard Trevithick, had shown that high-pressure steam engines could be made compact and powerful and portable.

It was Trevithick, remember, who drove a steam carriage through the streets of London in 1803, with eight none-too-sober cronies aboard—a steam carriage no one would put up money to develop. Perhaps Trevithick was a bit ahead of his time, for not many years later, in the 1820s and 1830s, there suddenly appeared a gaggle of as exciting and oddly contrived steam-powered road vehicles as ever turned a wheel.

Times had changed. The Napoleonic wars were over and England bloomed with prosperity. The new times and the new men, the spinners of Manchester, the iron founders, the potters, the gentry, were interested in new, faster ways of moving about the country. The miserable, potholed, gullied roads which had limited travel to horseback or slow, lumbering coaches in the eighteenth century were no longer good enough for the men of the Industrial Revolution.

Better roads were needed; roads upon which high-speed, four-horse coaches could dash across England with hurried stops at posting inns for fresh teams. And within a few years such roads were built using new methods devised and pioneered by those two great civil engineers, Telford and Macadam (whose name is still attached to a rather different road surface from the one he advocated).

The most successful steam carriages were built by Walter Hancock and Goldsworthy Gurney. Hancock built nine steam carriages with wonderfully peculiar names: *Autopsy, Era, Infant, Automaton,* etc. These machines, buses really, used steam at what was considered, in their day, exceedingly high pressure—200 pounds per square inch—but except in one instance were remarkably safe and reliable.

1. Cugnot's *Fardier* was a crude steam-driven three-wheeler for hauling artillery. This is the second version, built in 1771.
2. William James's steam carriage of 1829.
3. Cartoonist's version of a steam two-seater.
4. Drawing of Hills's steam carriage shows a huge vertical boiler but very little room for the engine attendant.

1

(Trouble befell the *Enterprise* when the engine attendant fastened the safety valve lever down with copper wire while the coach was standing still. Its boiler blew up.)

In 1834 *Era* and *Autopsy* carried four thousand passengers between the City and Paddington in London. Hancock's steamers were also rather more advanced in design than others of their time. They had wheel steering, their twin-cylindered engines were in separate interior compartments where they were protected from road dirt, and they drove through clutches and had chain drive. And Hancock's machines seem to have run at a profit.

Gurney built his first steam coach in 1827, and like other early locomotive builders, had qualms about the grip its wheels had upon the road. He therefore fitted it with a set of most ingeniously articulated legs and feet "to help when starting up and upon hills." This first Gurney bus was capable of carrying twenty-one passengers. It had a water-tube boiler and was fitted with separators to ensure dry steam for the cylinders. Londoners soon became inured to its noisy progress around town where it added its own small portion of smoke to that already smoky city, plus a not-too-welcome rain of cinders.

The passengers of steam carriages, especially those riding outside, also suffered from smoke and cinders and the discomforts engendered by iron-tired wheels, leaky cylinders, fragile boilers, and almost nonexistent brakes. Luckily for them few steamers could exceed fifteen miles an hour. At greater speeds the steamers would have been quite unmanageable, for they still had horse-vehicle steering in which the entire front axle turned upon a central pivot. The Ackermann steering motor vehicles use today, with each front wheel turning on its own pivot or kingpin, although known as early as 1714 and patented in England by Ackermann in 1818, wasn't used until 1873 on Frenchman Amédée Bollée's steam omnibus, *l'Obéissante*.

Not only mechanical problems bedeviled the hapless passenger. Filson Young in his book, *The Complete Motorist,* describes the hazards of one of Gurney's trial runs: "On one journey which he made to Bath with a number of guests, his carriage was attacked at Melksham where there happened to be a fair. The people formed such a dense mass that it was impossible to move the carriage through them; the crowd, being mainly composed of agricultural laborers, considered all machinery directly injurious to their interests, and with a cry of 'Down with all machinery' they set upon the carriage and its occupants, seriously injuring Mr. Gurney and his assistant engineer, who had to be taken to Bath in an unconscious condition."

The "assistant engineer" was what steam railroaders would later call the fireman. This poor man stood on the rear platform in some carriages, or in others inside among the hot furnaces, flailing rods, and leaping chains. His job was to shovel coal and coke (liquid petroleum fuels were some forty years in the future), grease the machinery with a brush dipped in lard or palm oil, and apprehensively check the safety valve to make sure that it wasn't stuck. Further, he had to maintain communication with the distant driver and slow or speed up the engine when so signaled. He in turn had to signal the driver when to stop for water, which was every ten miles or so.

Despite the steam carriages' shortcomings inventors and promoters rushed to build and operate them. Sir Charles Dance ran a line of Gurney steam coaches between Cheltenham and Gloucester from February to June, 1831, during which period he made 396 trips. Scott Russell operated a steam coach service between Glasgow and Paisley. Burstall & Hill, Macerone & Squire, James & Anderson are but a few of the steam-carriage builders whose names have come down to us.

We cannot divine whether the steam coaches would have replaced horse-drawn coaches, or whether small, individually owned steam cars would have been developed during those pre-Victorian days. For they were killed off by the combined opposition of the stagecoach operators, the turnpike owners who charged impossibly high tolls for steamers, and mostly by the new railways which, not too many years after Victoria started her reign, ended the day of the stagecoach.

Still, inventors made desultory attempts at building road steamers during the mid-nineteenth century. Thomas Rickett of Birmingham built a couple for very rich and eccentric milords just before the notorious "red flag" act of 1865 put

Patent Steam Carriage by W. H. James, Civil Engineer

2

3

LONDON
F. HILLS
PATENTEE

1

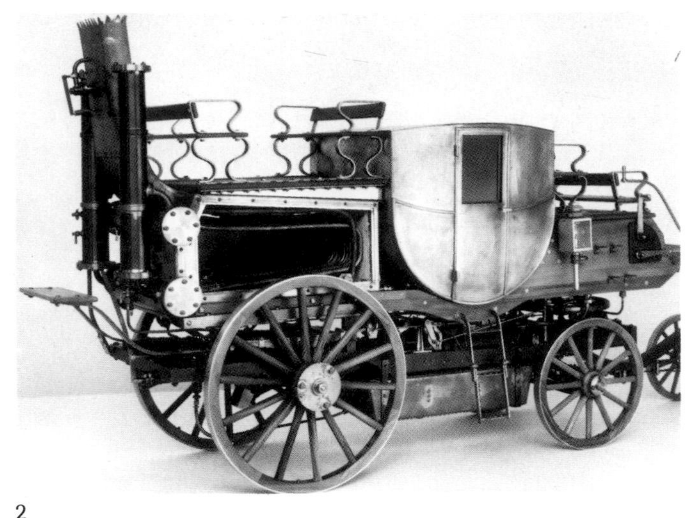

2

a stop to such playthings. This law made it illegal to drive any self-powered vehicle faster than four miles per hour in the country or two miles an hour in towns. And in either case a man had to walk ahead of the machine carrying a red flag.

In America similar road locomotives were constructed by Richard Dudgeon in 1853 and 1867. In 1869 Sylvester Roper of Roxbury attached steam engines to a velocipede and a buggy. Lucius Copeland built steam-engined tricycles and penny-farthing bicycles in the 1880s.

Now, too, another force was being developed which would make the automobile possible at exactly the time it was wanted, when the safety bicycle had created an appetite for independent travel. This was, of course, the internal-combustion engine.

The idea of causing a piston to move rapidly by creating a quick expansion of gas (an explosion) inside the cylinder containing it is as ancient as the stone cannon ball rushing through the barrel of a wooden cannon. Sir George Cayley, the father of aeronautics, actually built such a gunpowder engine in 1808 to power one of his flying machines. But gunpowder proved too violent and uncontrollable as a piston propellant.

Earlier, in 1784, an Englishman named Street had proposed an engine in which "the explosion was to be caused by vapourizing spirits of turpentine on a heated metal surface, mixing the vapour with air in a cylinder, firing the mixture and driving a piston by the explosion produced." Street was on the right track but seems never to have built his engine.

The first "practical" internal-combustion engine was built by Étienne Lenoir in 1860 and ran on illuminating gas. A Ruhmkorff coil supplied an electric spark for ignition. This Lenoir engine was as big as a bungalow and not much lighter. And it swallowed a hundred cubic feet of gas for each horsepower it developed! (The gas-air mixture, being uncompressed, pushed the piston with no great force when the spark ignited the charge.) Still, Lenoir succeeded in selling some of his remarkably inefficient engines to factories too small to afford steam-engine installations.

Lenoir experimented with liquid fuels, too, and in 1862 installed one of his engines in a crude, wagon-like vehicle which ran the six miles from Paris to Joinville-le-Pont in the unremarkable time of two hours. In the same year, Alphonse Beau de Rochas patented but did not construct a four-cycle engine.

It was Gottlieb Daimler, working in 1872 for the firm of Otto and Langen in Deutz, Germany, who first built a fairly successful four-cycle engine. Designed by Dr. Otto, it was a paragon of crudity and noise. The gas-air charge was fired to lift a monstrously heavy piston and toothed rod. When the piston descended on the next stroke it turned a flywheel by means of a rack to which the piston rod was geared—by gravity, in fact. Links and levers operated slide valves and a little slide which momentarily exposed a flame to the gas-air mixture. The racket perpetrated by this device was rather noisier than a thousand large beer steins falling down a flight of iron stairs. And about as reliable a source of power.

In 1876, however, Daimler succeeded in building what was called the "Otto Silent Gas Engine" which, if not too much more silent, was at last the direct ancestor of the engine in your car.

There was also another man who was involved with internal-combustion engines, perhaps even before Dr. Otto built his gas-powered noisemaker. He was that erratic dilettante, Siegfried Marcus of Vienna. In 1865 Marcus mounted a single-cylindered internal-combustion engine on a handcart, not to create a horseless carriage but to test his two-cycle benzene-fueled engine, which, since it had no compression stroke, was not much more efficient than Lenoir's huge monstrosity. The handcart had neither transmission nor clutch, and to start the engine it was necessary to lift the driving wheels off the ground and turn them by hand until the engine started. When the spinning wheels hit the ground again, the car, it was hoped, would take off. There is no real evidence that this brakeless device ever did more than limp noisily over the cobblestones of the *Mariahiferstrasse* in Vienna, where Marcus maintained his laboratory. But it evidently ran well enough to satisfy Marcus that he ought to quit fooling with it. For once he convinced himself that a thing worked he lost interest and headed off in nine other directions. In this case, however, he returned, some ten years later,

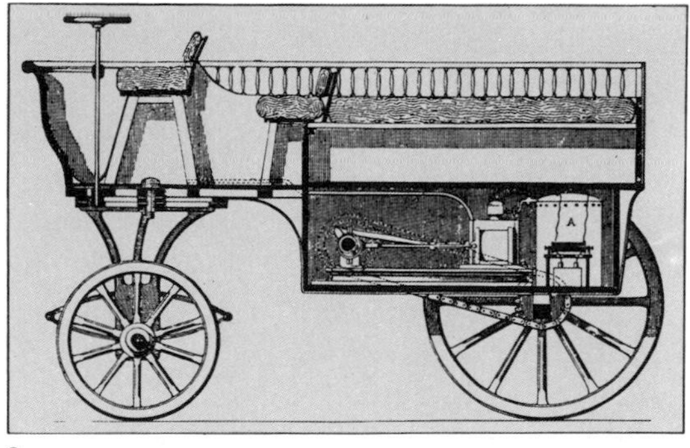

3

4

1. Goldsworthy Gurney was the most successful of steam-carriage builders.
2. Model of the Gurney carriage of 1827-28, with a side open to show boiler tubing.
3. Lenoir's gas-engined car of 1862. The engine (below) ran on illuminating gas.
4. Amédée Bollée's 1873 steam omnibus was one of the first vehicles to have Ackermann (Lankensperger) steering.

to his experiments with self-propelled vehicles. His *Strassenwagen* of about 1875 exists in the Vienna Technical Museum and will still run at about five miles per hour. Its chassis is of wood, its large wagon-like wheels have iron tires against which are pressed wooden blocks for braking—all very primitive.

But Marcus's engine was far more sophisticated than the rest of his *Strassenwagen.* Its single cylinder had a capacity of 1,570 cubic centimeters and developed about ¾ horsepower at 500 revolutions per minute. Marcus, whose specialty was electricity, designed a unique ignition system powered by a low-tension magneto. The carburetor, too, was a remarkably advanced mechanism, more so than those developed almost twenty years later. Exhaust gas was passed through its casing to warm the fuel-air mixture. The benzol was sprayed into a mixing chamber by means of a spinning brush which picked it up from a trough, and a hand-operated air valve controlled the strength of the benzol-air mixture. The intake valve was a sliding type like that on a steam engine; the exhaust left through a poppet valve. The car had a clutch and a differential, but Marcus seems not to have heard about Ackermann steering. The steering wheel turned the entire front axle, like that on a wagon. Again, Marcus became bored with his car. He did nothing to develop it further. It was a dead end.

Gottlieb Daimler was still working for Otto and Langen, who were still involved with selling their slow-moving, heavy gas engines when Marcus built his car. Daimler seems never to have known of Marcus's experiments. But he did know that he wanted to build a new kind of liquid-fueled engine that would be light and fast-revving (by nineteenth-century standards)—the kind of engine that could power boats, tramcars, horseless carriages, perhaps even cigar-shaped balloons.

Otto and Langen took a dim view of Daimler's ideas. Who needed a rapidly revolving liquid-fueled engine to power the belted and pullied power-transmission systems which were the normal means for running machines in factories? After a certain amount of fruitless hassling Daimler gave up trying to convince his employers. He struck out on his own in 1882, taking along Wilhelm Maybach, who had been closely associated with him. Within two years Daimler and Maybach had designed and built the first of the Daimler light four-stroke engines, and from this a still lighter engine evolved which was patented in 1885. It was a remarkably compact, high-speed (750 revolutions per minute) little power package, and except for its ignition system it was the direct forebear of the engines in the millions upon millions of motor cars which have been built in the last ninety years.

Nowadays an electric spark ignites the charge in a car's cylinders. Lenoir in 1862 had also used an electric spark, but his apparatus was big and clumsy. To nineteenth-century engineers (including, perhaps, Daimler) electricity was still a weird and incalculable force best left alone by solid citizens. Otto's engine, as we've seen, had used that open gas flame behind a little sliding door. But that was practicable only at very leisurely speeds. Daimler devised his own means of setting off the charge in his cylinder: hot-tube ignition.

Daimler's device was a small platinum tube which screwed into the side of the cylinder much like a modern spark plug. (Platinum was cheaper then!) A Bunsen burner, fed from its own little tank of benzene or gasoline, heated this tube red hot. When the rising piston compressed the mixture in the cylinder, some of it rushed into the tube and was ignited. Much simpler than a lot of coils, wires, and timers, but not really very good. First, before you could start cranking the engine you had to light the burner with a match, not easy on a windy day. And cars which later used Daimler hot-tube engines often had burners blow out when running into a wind. Further, such cars, if upset, often immolated themselves.

The first motor vehicle in which Daimler installed one of his engines was a bicycle of his own design—a primitive wooden motorcycle. He did this in 1885, when the first safety bicycles were appearing, and like them, his bike had two equal-sized wheels. But why he used a wooden frame and iron-shod wood-spoked wheels is something of a mystery. For by then steel tubing for frames and rubber-tired wire wheels were commonplace. Further, Daimler installed two small outrigger wheels, like the training wheels children use, and a leather saddle big enough to encompass a horse. The half-horsepower engine was installed in the same place motorcy-

2

1

cle engines have occupied ever since.

The motorcycle was no great success, although Daimler's son Paul managed to drive it the three kilometers from Canstatt to Unterturkheim and back.

Daimler was more interested in engines than in automobiles. As he had hoped, he installed them in boats, in tramcars, and even in a primitive dirigible. He was well aware that they would be wonderfully useful for the propulsion of road vehicles, but at first he looked upon them as accessories people might buy and install, instead of the horse, on the family carriage. He did so himself in 1886 and the first four-wheeled Daimler motorcar was no more than a horse-carriage with an engine poking up out of the floor in front of the rear seats. He did not use Ackermann steering; the front axle still swung on a center pivot. Still, this car ran satisfyingly, attaining some ten miles per hour from its single-cylinder 1½-hp engine. The rear wheels were belt-driven and the two speeds were selected by tightening and loosening the belts. There was no differential gearing between the rear wheels.

Although Daimler's primitive cars were in reality no more than test beds for his remarkable engines, he and Maybach quickly improved them and they became the precursors of that long line of motor cars which we now call Mercedes-Benz. But those wonderful little engines played a much more important role. Exported and licensed abroad (especially in France), they were the seeds from which most of the automobile industry grew.

At about the same time that Daimler's son was trying to tame his father's cranky wooden motorcycle, Karl Benz was trying out *his* first car, a spidery three-wheeler. Although Benz, like Daimler, was keenly interested in building small, light engines, engines per se were not his chief aim. Benz wanted mostly to build a self-propelled vehicle. Benz, too, had been a long-time expert in building stationary gas engines. But unlike Daimler, who had for years been comfortably important in his job with a stable and successful firm, Benz had lived on the very edge of poverty since boyhood. After years of struggle during which he did locksmithing, and worked in a railway engine shop and at various other mechanical pursuits, he turned to the production of gas engines. But he was continually plagued by lack of capital. Taking moneyed partners was no solution. Some cheated him. One pulled out his money and left Benz high and dry without even tools to work with. (They were sold to pay off the partner.)

Eventually he started another tiny operation with two friends, Max Rose and Friedrich Esslinger. This minuscule company, grandly named Benz und Cie., Rheinische Gasmotorenfabrik, at last freed him from many of his old troubles and plunged him into new troubles with his first motor car.

At first Benz und Cie. built stationary two-stroke engines with electric ignition designed by Benz. But he found that even the smallest of these was too clumsy and ran too slowly for the car he envisioned. Their top speed was a mere 120 revolutions per minute.

But what kind of car ought it to be? Benz had never seen any such thing. He therefore based his machine on the simplest light vehicle in common use, not the bicycle, which

1. Early stationary gas engines were huge, heavy, and slow-revving.
2. Siegfried Marcus built this *Strassenwagen* in 1875. It is now in the Vienna Technical Museum and is still capable of 5 mph.
3. Cross-section of the Phenix-Daimler engine.
4. Gottlieb Daimler.

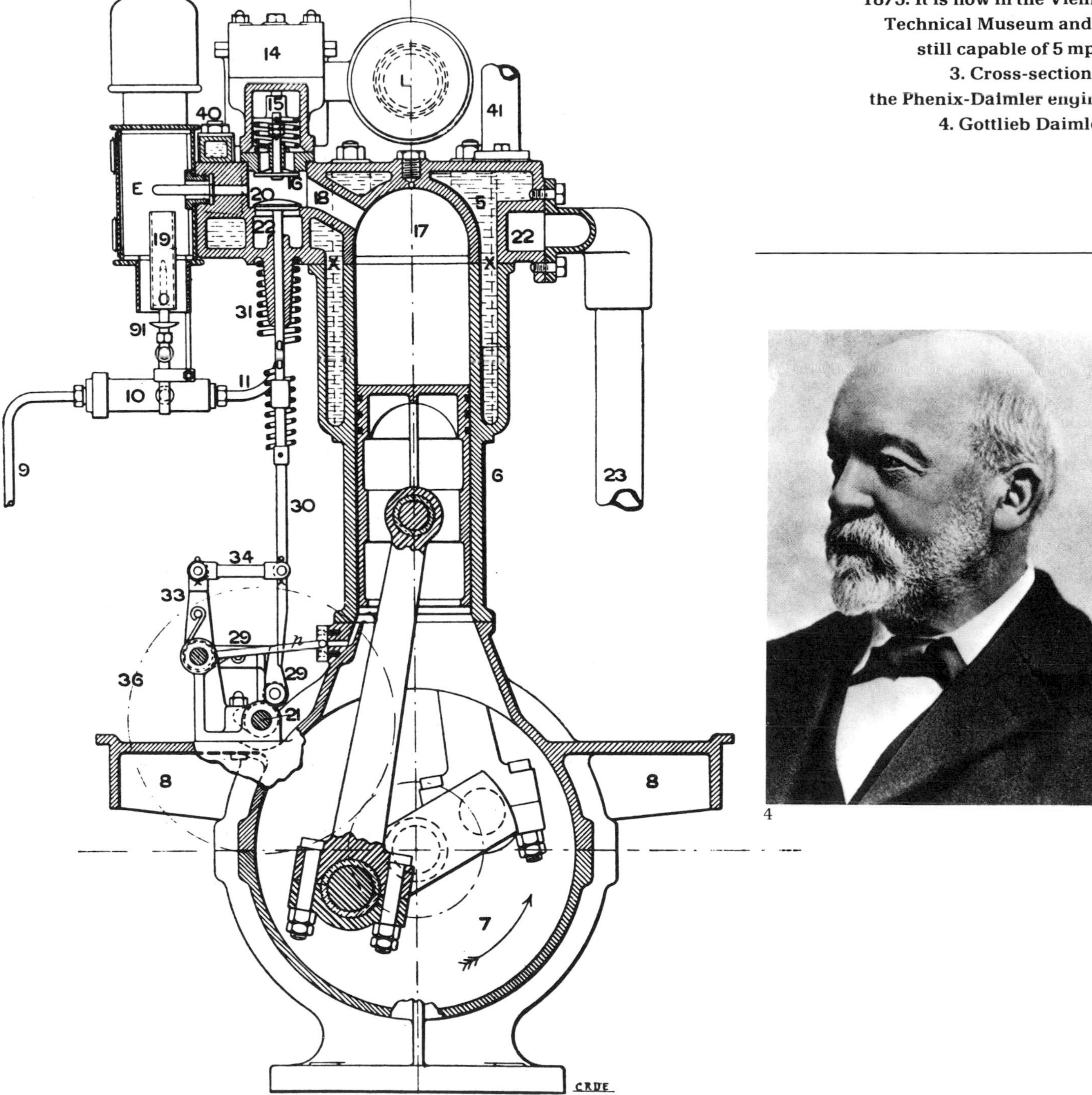

3

4

1

2

wouldn't stand up by itself, but on its sister machine, the tricycle, that quite ordinary device of the 1880s. Of course, Benz's machine differed from the pedal-propelled tricycle in several ways. It was somewhat heavier and bigger in order to sustain the weight and racking strains of an engine and transmission. And it also had to support the weight of a pair of proper seats rather than a saddle. One advantage of the tricycle configuration was the simplicity of its single-front-wheeled steering. Benz was perhaps ignorant of Ackermann steering. Or if he did know about it he was loath to involve himself in its complexities. And he rightly abandoned the archaic horse-carriage steering which at first satisfied Daimler.

Benz installed his single-cylinder, water-cooled engine in the rear of the machine. It developed ¾ horsepower at 250-300 rpm. Its connecting rod stuck nakedly out of the horizontally mounted cylinder and drove, by means of a similarly exposed crankshaft, a horizontal, spoked flywheel. No hand crank was needed. A belt led forward from the flywheel to a countershaft under the seats. And chains from the countershaft then drove the rear wheels. Oddly, only one speed was provided. But Benz *had* devised a differential gear.

Benz first drove his car for the delectation of his wife and children in the spring of 1885. It was autumn before he dared try it on a public road. Then, after being plagued by myriad troubles with the ignition and carburetion systems, he is said to have at last covered a kilometer at a speed of twelve kilometers per hour—an eight-minute run with time out for pushing.

To the detriment of his gas-engine business and the exasperation of his partners, Benz kept fussing and fiddling with his baby and constantly improving it. And running down the battery, too. Nights he'd lug the battery home and hook it up to a generator driven by the foot treadle of his wife's sewing machine. While Herr Benz sketched drawings for new improvements, Frau Benz pumped away to charge the battery for tomorrow's experiments.

It was Frau Benz, too, who went on one of the first long trips in a motor car. At five o'clock one morning, while Benz still lay asleep, she took her two older boys and set off in the car (a modified model) from Mannheim to Pforzheim. Eugen, the eldest, managed the tiller. On hills everybody pushed. But they made it. The first person to pilot a car on a longish road journey was a middle-aged German hausfrau!

The automobile was now almost ready for the world, but was the world ready for the automobile?

The world was indeed ready.

Since the end of the Napoleonic wars the people of Europe, at least those of the upper and middle classes, had felt ever more secure (only the minor Crimean war and that foolish adventure of Napoleon III, the Franco-Prussian war, had rippled the peace) and they had grown richer as the steam-powered factories had turned out an ever-increasing flood of goods which was sold not only to the rapidly burgeoning populations of England and Western Europe but also to the masses of colonials in India and Africa who labored to send raw materials to Britain and France and Germany in return. Profits were enormous.

The United States, too, was prosperous despite the trauma of the revolt of the slaveholding states in the 1860s. New land in the West and the tremendous growth of industry in the Northeast had seen to that.

By the 1880s also, people had long been used to mechanical transport. Train travel was commonplace. In the half-century since Stephenson's building of the Stockton & Darlington, railways had changed from a frightening new means of rapid travel to the normal way of going places. Every continent now had its network of tracks.

And people no longer thought mechanical things mysterious. By the 1880s thousands of factories wove cloth, turned and drilled metal, drew wire, milled grain, and performed the numerous tasks human and animal muscle had done a hundred years earlier.

But being rich and being used to mechanisms was not really enough to make people *want* automobiles. It was the bicycle which did that. It was the bicycle which gave people a taste of independent travel on the open road. And that taste of freedom aboard one's own wheels became an appetite which could not be denied. Daimler and Benz were the first to satisfy that appetite.

1. Daimler's first motor vehicle, a wooden motorcycle, built in 1885.
2. In 1886 Daimler installed an engine in a horse carriage.
3. Karl Benz. 4. Benz's motor-driven three-wheeler of 1885. The exposed engine at the rear was started by giving the flywheel a hefty pull.

3

4

5

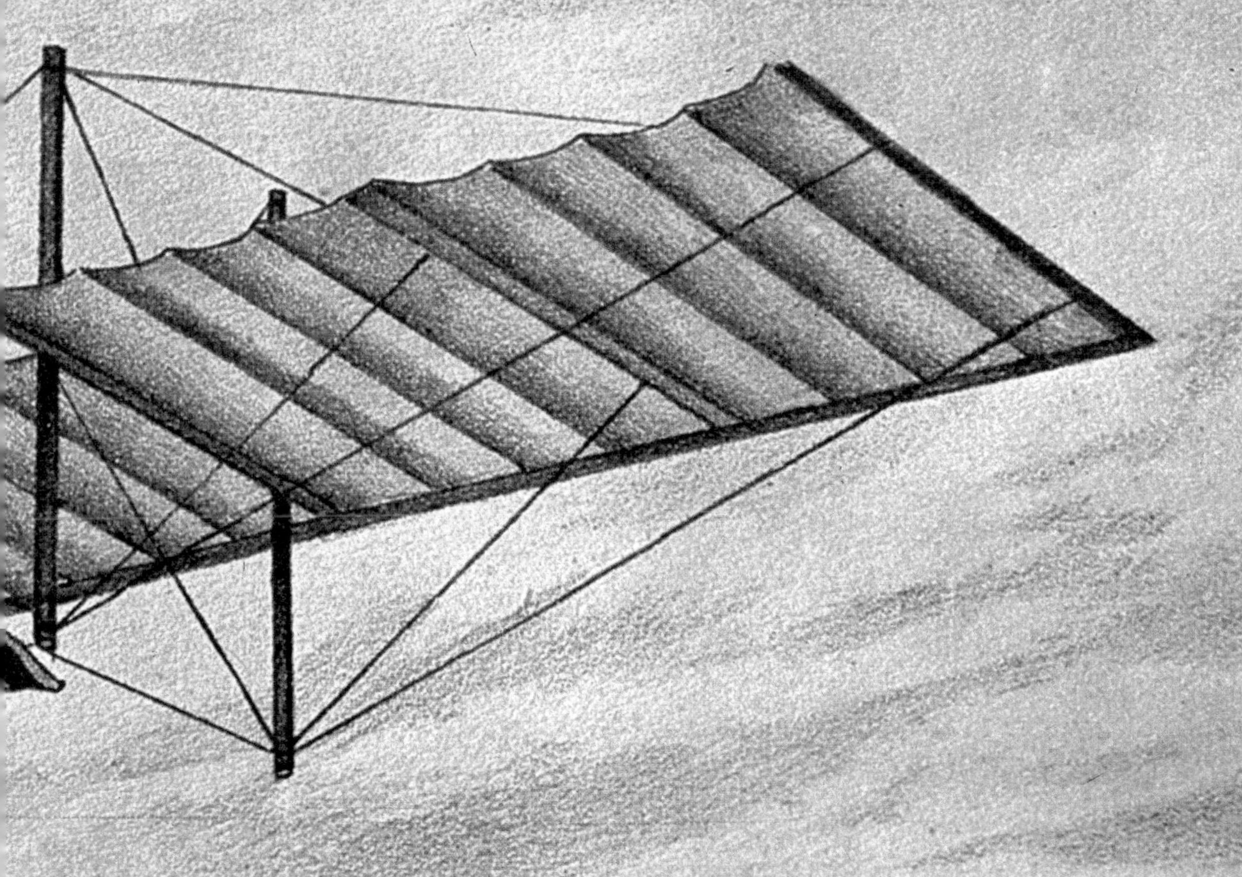

5 We Take to the Skies

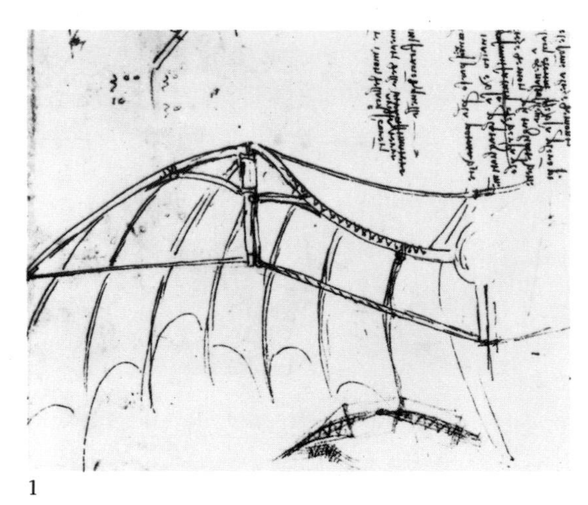

1

2

Earthbound No Longer

To fly like a bird. To soar freely, magically through the wide spaces of the air, escaping the earthward pull of gravity. That always has been the dream.

It has not been achieved. Mankind still cannot fly freely and quietly with the eagles. But millions of people do range the skies in great metal ships of the air, and a lot faster than any eagle.

If the planet was as big, say, as a basketball when it was first circumnavigated in Magellan's day by sailing vessels capable of perhaps five knots, the modern 1,200-knot-plus Concorde has reduced it to a pellet of birdshot. And the inhabitants of that shrunken world are, for better or worse, proportionately closer together.

It was in the late 1950s that jetliners began flying masses of tourists to places they previously had only read about. Today young people from Newark, New Jersey, go backpacking in Nepal, Japanese tour guides lead residents of Tokyo on canal trips in Amsterdam, factory workers from Manchester, England, loll on the beaches of Malta, and transatlantic flights are almost as numerous and crowded as commuter buses. People of every race, every nation, are learning about and perhaps accepting exotic cultures and customs. Will this make for more sympathetic understanding of each other's viewpoints and problems? We can hope so.

As far as is known, the first man-made device to take to the air was the balloon constructed and launched by the brothers Montgolfier. For eons before them, people had noticed that smoke and sparks rose upward from a fire. But it was these eighteenth-century Frenchmen who first thought of capturing whatever it was in the action of a fire that caused the smoke-and-spark ascension. They didn't know it was simply heated air.

They performed their first experiments in 1782, during a time of ferment such as the world has seldom seen. In America the revolution against the British Crown was, with French military help, almost over. In Britain the Industrial Revolution was accelerating. France, seven years distant from ridding itself of Louis XVI, would soon become embroiled with the rest of Europe in political and military upheavals that would not end until Waterloo.

The Montgolfiers, absorbed in "philosophic" experiments, as they were called in the eighteenth century, seemed unmindful of the chaotic world about them. Like other educated gentlemen of the time, they were in fashionable pursuit of the sciences. Ben Franklin, a companion spirit in Philadelphia, had taken some of the mystery out of electricity. Priestley in England and Lavoisier in Paris had identified oxygen, and Cavendish had isolated hydrogen. Some years earlier, in spite of king and church, Diderot had published his *Encyclopédie* with the help of forward-looking and rebellious intellectuals. Educated people were familiar with the works of Voltaire and Rousseau. Change was everywhere anticipated in science, politics, and the arts.

There was great excitement when the first balloons rose into the air. People were convinced that the Age of Flight was at hand. But ballooning was not flight. It would be more than a century before anyone would actually fly, rather than just support himself aloft on a bubble of gas. A heavier-than-air machine that could leave the ground, fly, and land safely could not be built until a number of scientific and technological advances had been made. Light and strong materials for constructing wings and air frames had to be developed, means of control in the air had to be invented, and aerodynamically correct wing and propeller shapes had to be determined.

Most important, a light engine, powerful for its weight, was required. It wasn't until 1885 that Gottfried Daimler broke away from the heavy, ponderous, slow-revving gas engines of the day and designed a light, comparatively high-speed gasoline engine. And it was the Wrights' development of such an engine that lifted Orville off the sands of Kitty Hawk.

The big airliners will fly ever faster in the future and, it is hoped, with less of the noise that has aroused the wrath of people living near jet airports. But the old dream of individual

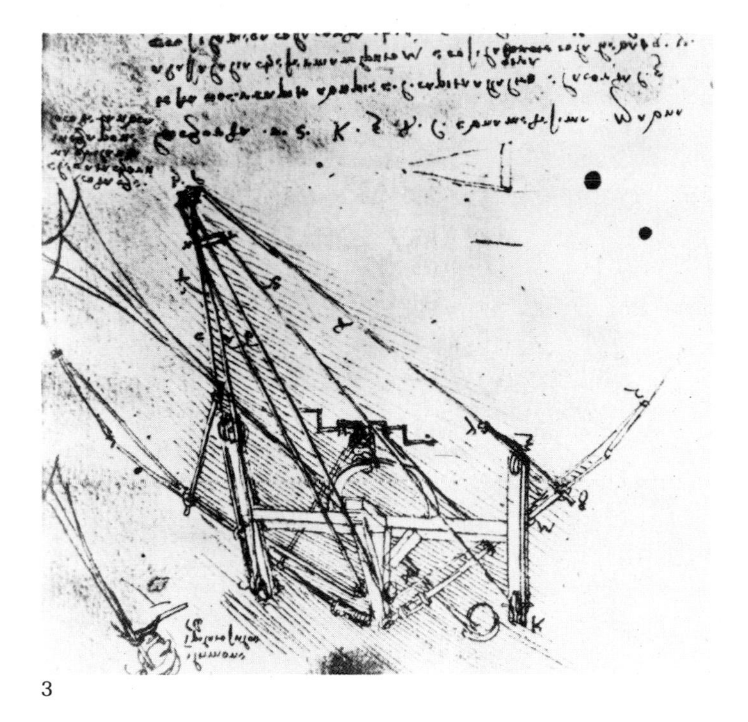

3

Preceding pages: Imaginary flight over London by W. S. Henson's "Aerial Steam Carriage" of 1842-43. 1. Drawing of an ornithopter by Leonardo da Vinci. 2. Self-portrait of Leonardo. 3. Leonardo's design for a man-operated ornithopter.

flight still eludes us. True, small aircraft are common. But even they require landing fields and a high degree of skill in their management. A flying device as simple and as personal as a bicycle still seems as remote as it was in Leonardo's day.

On a Bubble of Gas

Orville and Wilbur Wright were the first men on earth to fly, period. Certainly, other men before them had been lifted into the air. They had attached themselves to bubbles of hot air or hydrogen which, being lighter than ordinary air, rose as air bubbles rise in water. But these eighteenth-century balloonists did not, in the strictest sense, fly.

There always have been people whose envy of the birds has prodded them to devise means for flitting about in the air. No one knows how far back some humanoid tied some branches together, covered them with skins or leaves to make wings, leaped out of a tree and broke his hairy neck.

Much later, men knowing little or nothing of the complex structure of a bird's wing, and unaware that a bird's musculature is incomparably stronger in relation to its size and weight than a man's, persisted in trying to fly. Nor did these poor fellows know that a pigeon, for example, fills and empties its lungs some 400 times a minute while flying, or that a sparrow's heart beats 800 times a minute, these prodigious efforts supplying the blood with enough oxygen to maintain the strength of their muscles.

Even the great Leonardo da Vinci was kidding himself when he designed his lovely-looking winged mechanisms. His notion that a man using their ingenious levers and pulleys could amplify the strength of his arm, leg, and pectoral muscles sufficiently to fly by flapping strapped-on wings was faulty. As far as we know no one ever built one of Leonardo's flying devices. Leonardo also designed a model helicopter in about 1488. But this was little different from the helicopters which European children had had as toys for two hundred years, except that it used a continuous helical screw made of linen instead of individual rotor blades.

Generations of wing-flapping men continued to jump from towers and cliffs, doing themselves in thereby. But it was men of the other school of levitation—the gas-baggers—who first raised themselves aloft.

Roger Bacon, the medieval English monk who has been credited with all sorts of scientific feats—including that of being the first non-Chinese to discover gunpowder—wrote in his "Secrets of Art and Nature": "Such a machine must be a large hollow globe of copper or other suitable metal, wrought extremely thin in order to have it as light as possible, it must then be filled with ethereal air or liquid fire and launched from some elevated point into the atmosphere where it will float like a vessel on water." But just what "ethereal air" was he failed to say.

Soon afterward, a bishop of Chester refined Bacon's vague ideas by pointing out that the upper air was more rarefied and lighter, and that vessels filled with such low-density air would rise.

Various geniuses, noting that early morning dew rose off the grass when the sun's rays warmed it, had the brilliant thought that dew-filled egg shells (which were obviously thin and light) could be made to rise into the empyrean. All you had to do was collect enough dew and enough eggshells and wait for a hot sunny morning.

In 1670, twenty years after Otto von Guericke invented the vacuum pump, an Italian Jesuit, Francesco de Lana-Terzi, proposed filling four twenty-foot-diameter spheres of thinly beaten copper with nothing. The air inside them was to be sucked out by a pump. De Lana knew how much air weighed and worked out a means of descent by the admission of air into the spheres. He didn't realize, however, that atmospheric pressure would instantly squash his thin spheres. Anyhow, de Lana feared that the intercession of God would prevent the flight of his machine "since it would cause much disturbance among the civil and political governments of mankind."

Another Jesuit, the Brazilian Laurenço de Gusmão, actually seems to have demonstrated in 1709 that a miniature

1. Airship designed by de Lana in 1670.
2. Montgolfier hot-air balloon, which made its ascent at Lyons on January 19, 1784.
3. Trial of a *Montgolfière* in 1783.
4. German version of the world's first aerial voyage. Pilâtre de Rozier and the Marquis d'Arlandes made a five-mile trip over Paris in a *Montgolfière* on November 21, 1783. This artist's *Montgolfière* is not quite correct.

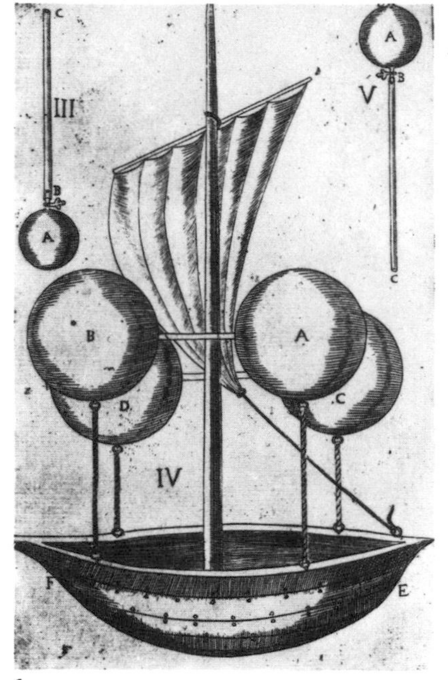

1

2

3

A
B
E
F
F
D

Opposite: The first hydrogen balloon, a *Charlière,* rose from the Tuileries Gardens in Paris on December 1, 1783. It was called a *Charlière* after J. A. C. Charles, who designed it and worked out a method of generating hydrogen to fill it. Aboard with Charles was Monsieur Robert, who helped construct the balloon.

hot-air balloon could fly. He showed up with his device at the Portuguese royal palace and proceeded very nearly to burn the place down. According to a manuscript found by Charles H. Gibbs-Smith, the noted British aeronautical historian, the model balloon "consisted of a small bark in the form of a trough which was covered with a cloth of canvas. With various spirits, quintessences and other ingredients he put a light beneath it, and let the said bark fly in the Saila das Embaixadas before His Majesty and many other persons. It rose to a small height against the wall and then came to earth and caught fire when the materials became jumbled together. In descending and falling downwards it set fire to some hangings and everything against which it knocked. His Majesty was good enough not to take it ill."

It was hydrogen which made ballooning practical—if balloons and dirigibles are ever really practical. In 1766 the English chemist Henry Cavendish isolated hydrogen—"inflammable air," as it was then called—and reported that it was much lighter than plain old air. Several scientists considered building hydrogen-filled balloons, but before anything was accomplished a couple of Frenchmen became the first men to leave the earth.

Joseph and Étienne Montgolfier were prosperous papermakers in the town of Annonay, near the city of Lyons. In the late eighteenth century, it was expected that well-educated, prosperous young gentlemen would indulge in the new hobby of science. Electrical and chemical experiments were performed in drawing rooms, and it was indoors, in 1782, that Joseph first tried to capture the strange "gas" which propelled sparks and smoke up the fireplace chimney.

He had his landlady make him a silk bag open at the bottom. When he held the bag over a fire it inflated. When he let go of it, it flew up and rested against the ceiling, frightening the poor landlady. Joseph wasn't at home when he performed his experiment. He was away in Avignon, and he wrote Étienne, "Prepare promptly a supply of taffeta and ropes and you will see one of the most astonishing things in the world!"

The silk bag went up more than seventy feet when the brothers tried it outdoors. They didn't tell anybody about what they were doing but kept on making ever bigger "aerostatic machines." Using a fire kindled of chopped wool and straw which made a lot of black smoke and which they thought contained a mysterious new gas, they succeeded in raising a balloon to an altitude of more than a thousand feet before its air cooled and it came down, a mile from where it had been released. The brothers still had no inkling that it was hot air rather than a fancy new gas which was lifting their balloons. They even had a sneaking suspicion that electricity, that recently discovered new force, might somehow be responsible.

In the spring of 1783, in Annonay's market square, the Montgolfiers put on a public show of their "aerostat." Surrounded by "a respectable assembly and a great multitude of people," they kindled a fire under a linen bag lined with paper. It was a good-size envelope over a hundred feet in circumference, and when it was inflated with hot air it required the efforts of eight brawny peasants to prevent its taking off. The crowd was astounded. But it was even more astonished when, on an order from Joseph, the world's first ground crew let go, and the aerostat bounded aloft, rising six thousand feet. "Philosophical instruments" measured the altitude. The balloon then slowly descended, landing about a mile away.

The news of the unprecedented flight caused wild excitement among the members of the Academy of Sciences in Paris, and the Montgolfiers were summoned to demonstrate their aerostat to the savants on the Seine. The youthful "philosophe" J. A. C. Charles was also asked to further delve into the mysteries of aerostation.

Monsieur Charles, who was aware of Henry Cavendish's hydrogen-gas experiments, doubted that whatever it was that was levitating the Montgolfiers' device had the lifting power of Cavendish's "inflammable air." He decided to use the gas instead. Concerned lest the hydrogen leak through the pores of the paper and linen, he asked a pair of brothers, the Roberts, who had recently perfected a method of coating silk with rubber, to make a "flying globe" of their new impermeable fabric.

The little thirteen-foot-diameter balloon soon was ready, but Charles had trouble generating enough hydrogen

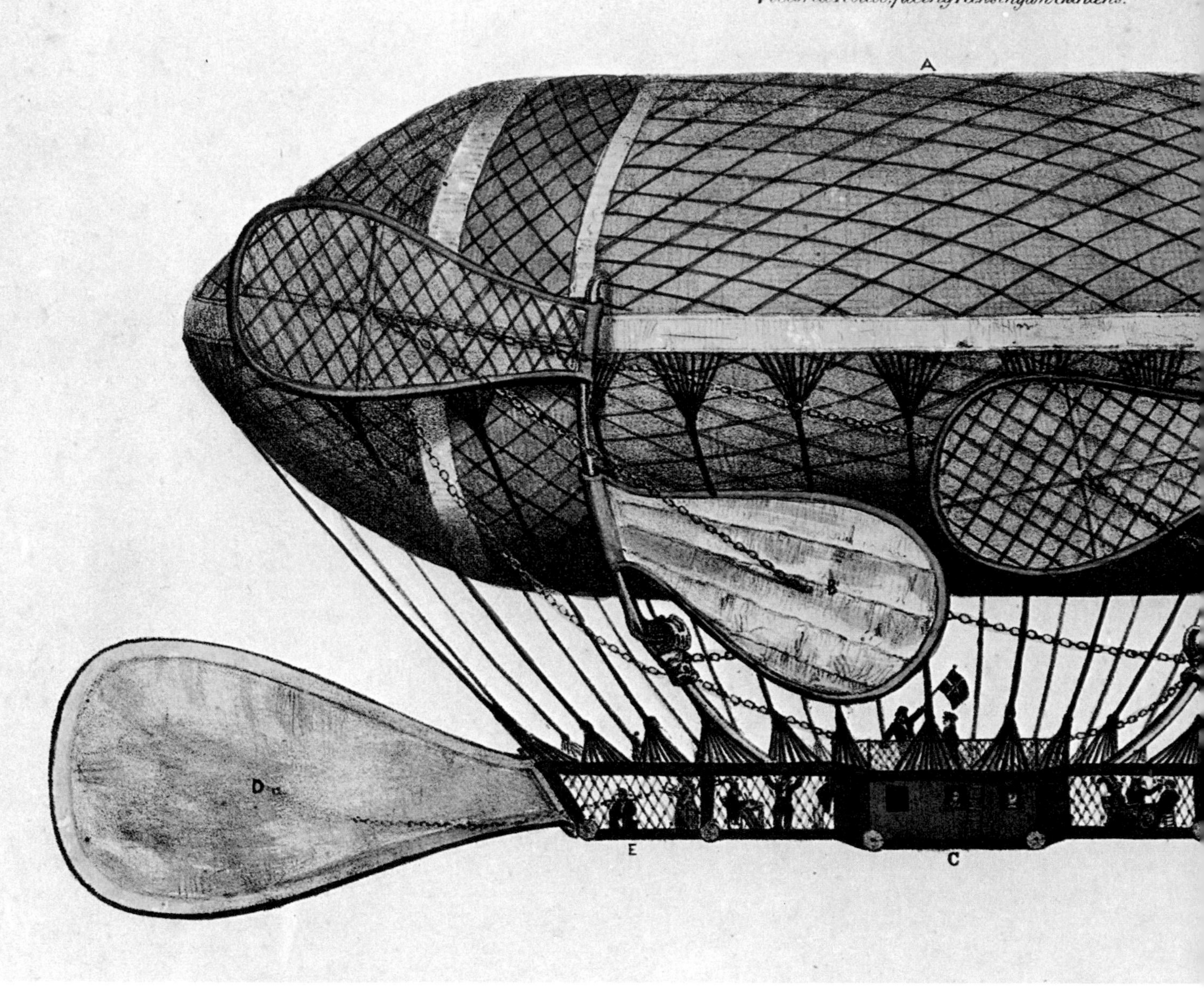

to fill out its wrinkles. Several days later, although the balloon still was not fully inflated, Charles judged that it had enough gas to be moved to the Champ de Mars (where the Eiffel Tower now stands) for its ascension. All Paris was by now in such a state of excitement that Charles thought it best to avoid the crazy crowds by transporting the balloon in the middle of the night.

But the dark of night didn't faze the people of Paris. Knowing nothing about the explosive properties of hydrogen, thousands of them carrying flaming torches followed the balloon. Luckily neither the "flying globe" nor Charles's generating apparatus blew up.

From daybreak until five o'clock in the afternoon Charles made desperate efforts to fill his balloon. By then it was raining, but he launched it into the low gray clouds, anyhow. Benjamin Franklin was in the huge neck-craning crowd. "What use is it?" someone asked him. "What use is a newborn baby?" said the eighty-one-year-old electrical experimenter and Minister Plenipotentiary.

Étienne Montgolfier, meanwhile, had built one of his paper-and-linen hot-air "aerostats" for his demonstration before the Academy. It was a giant, seventy-four feet high, and fancifully painted. He did so well with it that he was asked to come to Versailles to show his amazing device to Louis XVI and Marie Antoinette. But Étienne had to put a new balloon together. The one the academicians had seen had been out in the rain, which didn't do its paper and linen much good. Étienne whipped out a new one in four days.

This time a rooster, a sheep, and a duck went up in the balloon. The Montgolfiers wanted to find out if life could be sustained at high altitude.

Eight minutes later and a mile and a half away the *Montgolfière* landed. (Hydrogen balloons were called *Charlières.*) The flying zoo was alive and well. And the king was happy. The brothers got the Order of St. Michel, and Louis ordered a gold medal struck in their honor "*pour avoir rendu l'air navigable.*"

The first man to ascend in a balloon was a young physician, Jean-François Pilâtre de Rozier. And two years later he was the world's first man to die in a flying accident.

In their efforts to control the direction of balloons, designers dreamed up such devices as the man-powered paddles on the Comte de Lennox's proposed 1835 airship *Eagle* (left), and the rudimentary airscrews on the lentil-shaped Italian balloon below, shown in the air near Bologna in 1838.

Pilâtre had been captivated by the idea of being a balloonist since the moment he saw the *Montgolfière's* menagerie touch down safely. He pestered the Montgolfiers with such persistence that when they built a third balloon he was one of the two men who went up in it as a fire tender and stoker. They rode, not in a basket, as on a modern balloon,but on a sort of gallery made of wicker which encircled the fire, a very risky business indeed, considering the fact that the flame was kindled under a paper-and-linen structure.

After several ascensions tethered at the end of a rope, the *Montgolfière* made its first free ascension in November, 1783, carrying Pilâtre and the Marquis d'Arlandes. For some twenty-five minutes they sailed five miles or so over Paris, waving their cocked hats at the thousands of upturned faces below when they weren't busy slapping at small fires started by sparks.

Almost immediately after Pilâtre's success aboard the *Montgolfière,* Professor Charles made the world's first ascent in a hydrogen-filled *Charlière.* He took along one of the Robert brothers, food, a barometer and other instruments, and bags of sand as ballast which could be jettisoned to gain altitude. For the first time, too, Charles suffered from a new malaise: the painful effects of altitude. At nine thousand feet, cold and pains in his ears forced him to descend.

It is difficult to realize what a tremendous sensation these first essays at leaving the earth's surface created. A ballooning craze gripped the civilized world. But some people worried. Horace Walpole wrote: "I hope these new mechanic meteors will prove only playthings for the learned and the idle, and not be converted into new engines of destruction to the human race."

Ascents were made in England and in the United States. An Englishwoman, Mrs. Letitia Sage, said to have been remarkably beautiful despite a weight of more than two hundred pounds, was the first person to cause the dumping of a passenger on a flight. In 1785, when Vincent Lunardi, George Biggin, and Mrs. Sage were to take off from London, Mrs. Sage's weight was too much for the *Charlière.* Gallantly, Biggin got out.

In the same year the Channel was crossed from

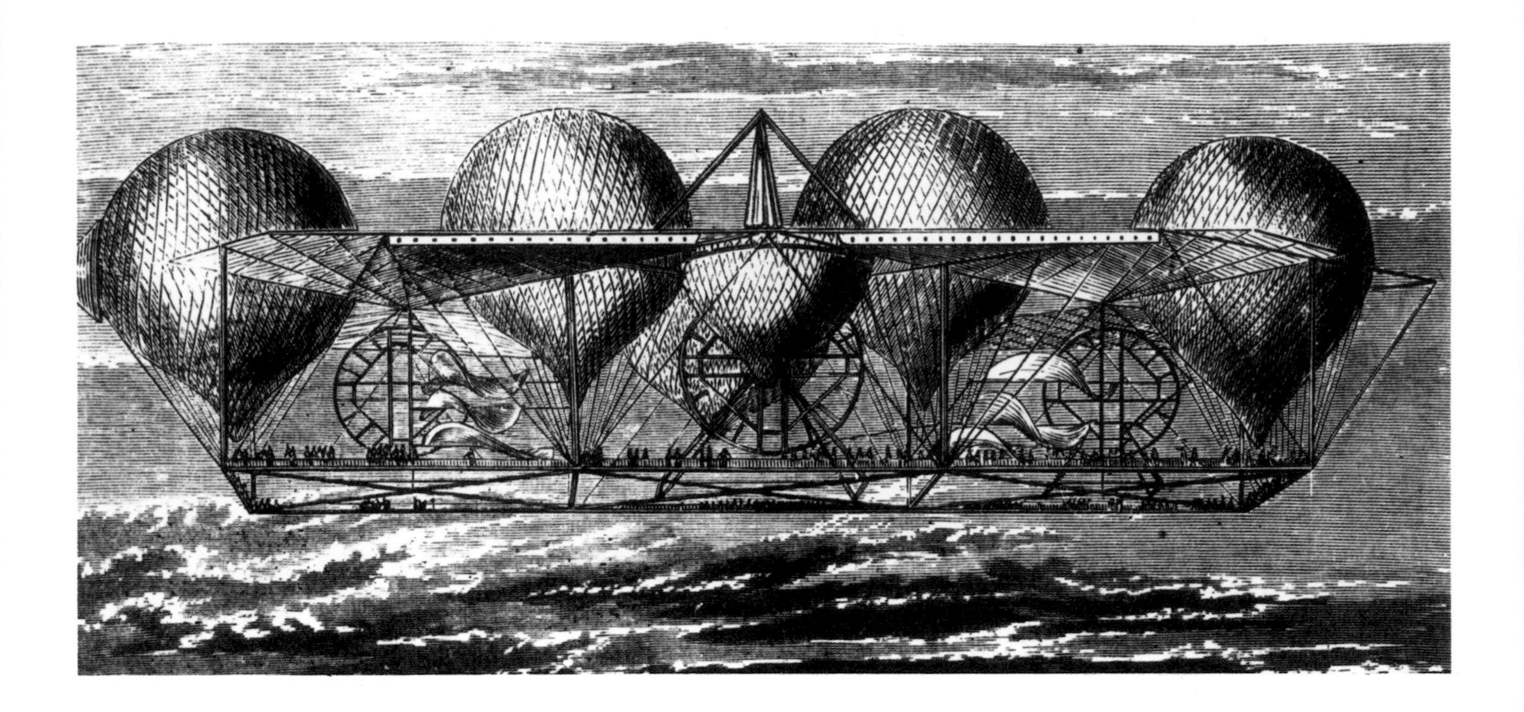

England to France, while Pilâtre de Rozier suffered the first aerial disaster while trying a crossing from France to England. Amazingly, no one until then had been injured while ballooning. Pilâtre, with incredible foolhardiness and with financial help from the French government, whose ignorance may be excused, had built a balloon which was not only filled with hydrogen, but also had a cylindrical *Montgolfière* complete with its bonfire below it. Professor Charles warned him that he was mixing fire with gunpowder. Still, with a passenger—Pierre Romain, who had manufactured the balloon—he took off from Boulogne. After it rose about three thousand feet spectators on the coast saw a puff of blue flame, and heard the thump of an explosion. Pilâtre and Romain both were killed.

Although the frenzied enthusiasm for ballooning subsided somewhat in the nineteenth century, balloons continued to be attractions at fairs, and were used to a limited extent by the military for high-elevation reconnaissance during the American Civil War. A number of long-distance flights were attempted. An effort to cross the Atlantic in 1859 failed. In 1897 Salomon Andrée tried to reach the North Pole. The remains and undeveloped photographs of that expedition were found in 1930. (The films were successfully developed.)

Unpowered spherical balloons can do no more than ride the wind. They cannot be steered. The next step for inventors was the powered, steerable balloon. In the early days of ballooning, steam engines were still huge, heavy, and obviously impractical, and a usable internal-combustion engine was still a century away.

Still, the first inventors who tried to build balloons which would go where you pointed them realized that they ought to be elongated, cigar-shaped. One of the first such "dirigibles" was designed by a French physicist and built by the brothers Robert in 1784. Five parasol-shaped oars were used in a vain effort to propel it. Later inventors proposed paddle wheels, flapping wings, captive eagles, even propellers. But since all of these—except the birds—had to be worked by hand they failed.

In 1843 an Englishman, Monck Mason, succeeded in moving short distances in a forty-four-foot-long dirigible

Opposite, top: Multiballooned airship, the brainchild of a Monsieur Pétin in 1850. Bottom: Gaston Tissandier dropping a handful of sand ballast from a balloon in 1887. Left: Filling a nineteenth-century balloon with hydrogen.

powered by wind-up clockwork which turned a propeller.

It was Frenchman Henri Giffard who first used mechanical power—steam power—to push a cigar-shaped gas bag through the air. Pictures of Giffard show him standing top-hatted on a little open platform slung a long distance below his forty-four-meter balloon. On the platform with him is a cylindrical coke-fired furnace and boiler, and a small three-horsepower steam engine which drives a flat-bladed propeller. Smoke pours out of a skinny funnel. Rigged to the complex cordage attaching the "car" to the balloon is a huge, sail-like rudder, which can be turned by means of a rope. Giffard managed to steam along at about 5 mph for seventeen miles before landing. But steam engines were too heavy and the idea was abandoned despite later experiments by Giffard. In 1872 another Gaul, Henri Dupuy de Lôme, reverted to manpower to turn the propeller on his dirigible. Eight men turning a long crank built into a boat-shaped wicker basket failed to supply enough power.

In Austria, Paul Haenlein, a German, used a Lenoir-type gas engine to propel a quite modern-looking dirigible. The four-cylinder engine, its radiator, and seventy-five liters of water weighed about a thousand pounds and proved to be too heavy. Ingeniously, Haenlein used hydrogen from his gas bag for fuel.

Electric motors fed by storage batteries were tried next. In 1883 Gaston and Albert Tissandier made a five-minute flight over the Bois de Boulogne, near Paris. Later they succeeded in a flight across Paris. But their heavy electric motor, plus twenty-four even heavier Planté batteries, were still not the answer. Paul Renard and Arthur Krebs were somewhat more successful in their use of electricity. In 1884, aboard the 160-foot dirigible *La France,* they succeeded in making the world's first aerial journey which ended at its starting point, near Meudon. They covered five miles.

Gasoline engines had been in use for years before anyone tried to power a dirigible with one. It wasn't until 1898 in Paris that the flamboyant and very rich little Brazilian, Alberto Santos-Dumont, hung a modified version of a De Dion-Bouton motor on a rather floppy dirigible. Later Santos-Dumont built a long series of dirigibles (sixteen of them) with ever bigger engines and became the darling of the Parisians as he survived one comic accident after another. His machines collapsed, hit trees, were destroyed by hoodlums, and once he claimed that the wind from his propeller gave him pneumonia. At least once, wearing his usual impossibly high collar and trick Panama hat, he flew model No. 9 to his elegant town house, where footmen, chauffeurs, and gardeners caught hold of its trailing ropes, hauled it down, and held it while the little dandy took some refreshment. In 1901 Santos-Dumont became world famous by flying in his No. 6, from St. Cloud, around the Eiffel Tower and back in twenty-nine minutes and thirty-one seconds.

The marriage of the internal-combustion engine and the elongated gas bag at last made the dirigible seem practical. In France, the 190-foot *Lebaudy* was making long trips by 1902. But it was the German Count Ferdinand von Zeppelin's huge, metal-framed, rigid airships—the first one built in 1900—which gave the world the illusion that a safe, luxurious, fast means of flying transport was at hand. That illusion was shattered in 1936, when the *Hindenburg* exploded at Lakehurst, New Jersey.

With Wings and Pistons

In 1903 the Wright brothers were the first to *fly* an airplane, but Sir George Cayley had *invented* the airplane a century before they left the ground at Kitty Hawk, North Carolina. In 1804 Cayley flew a small model whose wing consisted of a kite, and whose fuselage was a five-foot-long stick at the rear of which was an adjustable tail unit incorporating the means for steering and vertical control. Five years earlier, in 1799, he had scratched a small silver disc with a diagram showing three forces which affect the action of a flying wing: thrust, lift, and drag. On the obverse side he showed a fixed-wing airplane, with vertical and horizontal tail surfaces and paddles for propulsion.

1. & 2. Silver disc (obverse and reverse) upon which Sir George Cayley, in 1799, scratched diagrams showing the principles of flight.
3. 1819 sketch by Cayley of a model glider, showing the rudder on the elevator and the wing curvature caused by airflow.
4. 1853 sketch of an improved version of Cayley's glider.
5. Sir George Cayley.

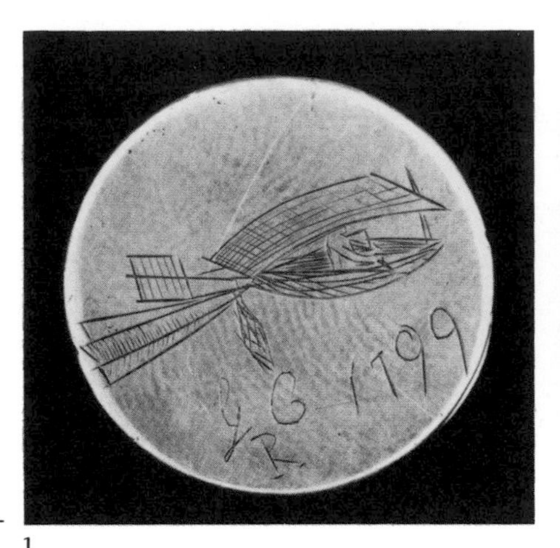

1

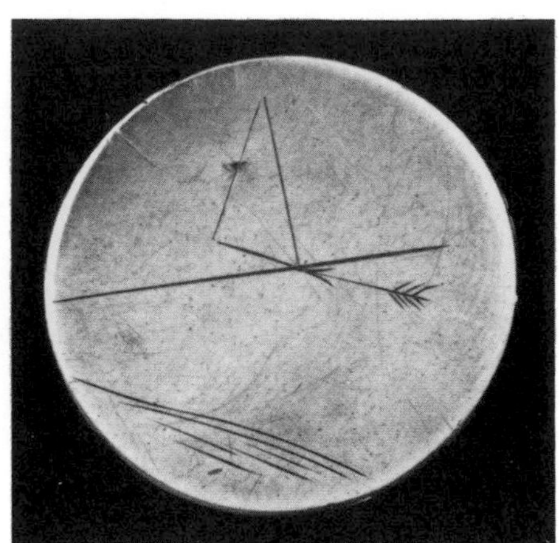

2

I have in front of me a pamphlet, "Aerial Navigation," which Sir George wrote in 1809. Yes, 1809! In it he set forth the principles on which later investigations of aerodynamics were based. It is hard to realize that even before railroads or steamboats were in use Cayley wrote: "We shall be able to transport ourselves and families, and their goods and chattels, more securely by air than by water, and with a velocity of from 20 to 100 miles per hour. To produce this effect it is only necessary to have a first mover, which will generate more power in a given time, in proportion to its weight, than the animal system of muscles."

Amazingly, he foresaw the use of the gas engine: "Lightness is of so much value in this instance that it is proper to notice the probability that exists of using the expansion of air by the sudden combustion of inflammable powders or fluids . . . probably [an] engine of this sort might be produced by a gas-light apparatus and by firing the inflammable air generated with a due portion of common air under a piston." Fifty-six years later Lenoir built such a "gas-light" engine, which soon was developed by Otto and the rest into the modern gasoline engine.

George Cayley was ten years old in 1783, when the Montgolfier brothers took to the air in their wonderful paper bag full of hot air. And like so many small boys, he started making his own hot-air balloons using stubs of candle and tissue paper. (On the Fourth of July in my youth kids in the Bronx were still flying such hot-air devices and keeping the Fire Department busy putting out fires on roofs.)

George Cayley devoted his life to the study of air, and birds in that air. He learned about wing shapes, about air density. He built a whirling-arm apparatus to study the action of various wing shapes in motion through the air. He found out that curved wings had more lift than flat ones. He proposed that wings be superposed; a biplane, he realized, would give more lift with less weight. And he showed that setting the wings at a dihedral (a flat V) would improve stability.

Cayley built gliders, monoplanes, biplanes, and triplanes that incorporated his discoveries. In 1849 one of them lifted a ten-year-old boy "off the ground for several yards on descending a hill." In 1853, when Cayley was eighty years old, he talked a less-than-happy coachman aboard one of his gliders and pushed him off a hill. The poor coachman sailed across a valley and landed in a cloud of dust.

Wrathful, he extricated himself from the glider and quit on the spot, saying,"Please, Sir George, I wish to give notice. I was hired to drive, and not to fly!"

Victorian England lived by steam. Steam powered everything that moved: railways, ships, machines in factories. It was natural then that even when a man wanted to spin the propellers of an aeroplane he turned to the steam engine, hoping that somehow, someone would magically provide him with a very light one.

The aeroplane William Samuel Henson designed in 1841 was, as might be expected, steam-engined. Henson was very familiar with Sir George's work. It was he who had dubbed him "father of aerial navigation."

Henson's "Aerial Steam Carriage" is to our eyes a very modern-looking flying machine, much more modern looking, really, than the kite-like contraption the Wright Brothers flew sixty years later. Its 150-foot-wide wing was double-surfaced, cambered, and ribbed, and would have looked perfectly correct on a 1930s machine. It had a tricycle landing gear and a cabin which the passengers would share with the steam engine that drove twin, six-bladed pusher propellers. And it set the fashion for the way people thought a proper aeroplane ought to look. For its day Henson's "Aerial Steam Carriage" received a tremendous amount of publicity. Henson had filed beautifully rendered drawings when he applied for a patent in 1843 and fanciful illustrations of his aerial steam carriage, the *Ariel,* appeared in magazines, as prints, and even as pillow covers and handkerchiefs. It was shown flying over London and the pyramids at Gizeh, among other places. Henson's machine was never even built, let alone flown, and his grandiose scheme to form an Aerial Transit Company came to nought after a large model he constructed with the help of a friend, the engineer John Stringfellow, failed to do more than glide downward after leaving its launching ramp. Henson gave up, but Stringfellow continued the experiments and built a model monoplane in 1848. Despite its light and elegant steam engine and boiler, it too

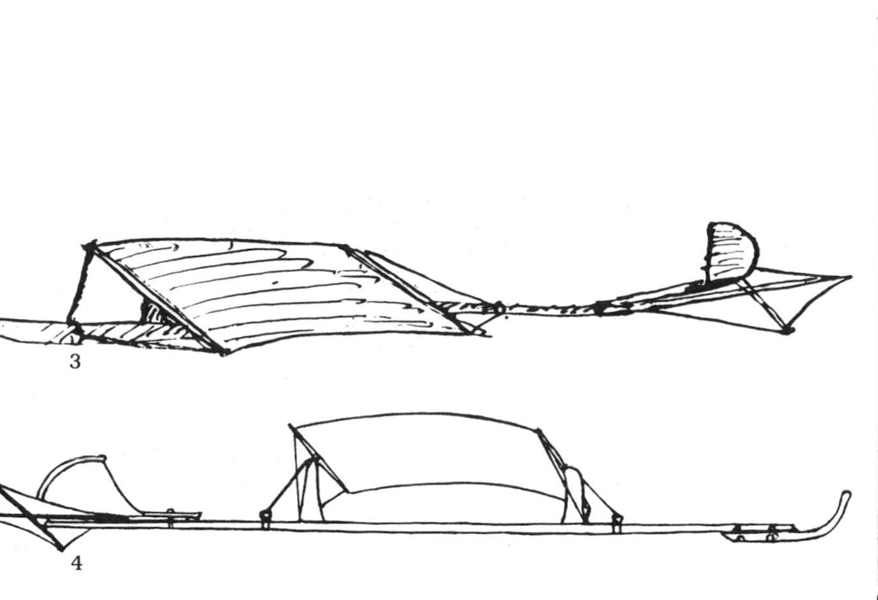
3

4

5

failed. Stringfellow showed a model triplane in 1868 at the exhibition of the Aeronautical Society of Great Britain (now the Royal Aeronautical Society) at London's Crystal Palace. But this twelve-pound model powered by a one-third horsepower steam engine also did not sustain itself in the air. Still, both Henson's and Stringfellow's designs had a notable influence on the configuration of later aircraft. Monoplanes built a hundred years after Henson's day still looked like his aerial steam carriage, and almost every biplane or triplane ever built had superposed wings much like Stringfellow's model.

It was a paper read by another Britisher, F. H. Wenham, at the first meeting of the Aeronautical Society in 1866 that had inspired Stringfellow to build his triplane model. Wenham's experiments in aerodynamics had convinced him that a long skinny wing had more lift than a short squarish one. But a long narrow wing was difficult to brace; it tended to buckle. He therefore proposed that wings be superposed for greater strength. In the world's first wind-tunnel tests, Wenham also corroborated Cayley's theory of 1809: that the partial vacuum above an aeroplane's arched wing does more to sustain it in the air than does the pressure of the air beneath it, a fact not fully proved until a half-century later, when aircraft actually started flying.

In the 1870s a young French genius, Alphonse Penaud, succeeded in solving many of the problems of flight—all except that of motive power. He built lovely, stable models powered by rubber bands: helicopters, wing-flapping ornithopters, and, most promisingly, aeroplanes, which he called *planophores*. One of his *planophores* flew more than 125 feet in eleven seconds.

His design for an amphibian, which he patented in 1876, still looks practical. It had a wing shape not unlike that of some jet fighter planes, a retractable landing gear, a transparent cockpit canopy, and a joystick to control rudder and elevators. It lacked two important features: ailerons and an engine to turn its twin propellers.

Discouraged by his peers and the pompous pundits of the French scientific establishment who ridiculed him, Penaud shot and killed himself in 1880 at the age of thirty.

The first full-sized flying machine to hop off level ground was Clément Ader's *Eole*, at Armainvilliers in 1890. But it was just that—a hop, not a flight. Although the internal-combustion gasoline engine was available by 1890—in a primitive stage of development—Ader still opted for the steam engine. *Eole* had bat-like wings, and a 20-hp engine which drove a four-bladed propeller. Ader built two other machines. He gave up on the second without attempting to fly it. His third, *Avion III,* which he built with some financial help from the French army, had twin steam engines and two propellers. This one didn't even hop when it was twice tested at Satory in 1897 before official witnesses. The French military withdrew its backing and Ader abandoned the project. Yet, in 1906, after the success of the Wright brothers, Ader insisted that he had flown three hundred meters in 1897. Although it is obvious that official observers saw no such flight, Ader still has chauvinistic French supporters who insist that he flew six years before the Wrights. You can still see Clément Ader's remarkable-looking machine hanging from the rafters of the Conservatoire National des Arts et Métiers in Paris.

Another fantastic nonflying aircraft of the 1890s was that built by the bearded little transplanted American, Sir Hiram Maxim. Pompous and noisy, like more than a few men of short stature, Maxim was well-heeled and well-known mostly for having invented the machine gun which bore his name.

Maxim was interested in aeronautical experiments to prove flight was possible. He had no urge to leave the ground in his monstrous machine. He tried a variety of wing shapes and experimented with a number of propellers. Then he built a half-mile of nine-foot-gauge railroad track on his estate southeast of London. On it he mounted his gargantuan machine. It had a wingspan of 104 feet and had its wings, some of them very oddly shaped indeed, set in fantastic array. The total lifting surface was six thousand square feet. A forest of steel struts and wires held the thing together. Twin compound steam engines with a then-unbelievable 360 horsepower turned a pair of 17-foot, 10-inch propellers. With six hundred pounds of water in its tanks and boilers, and with

1

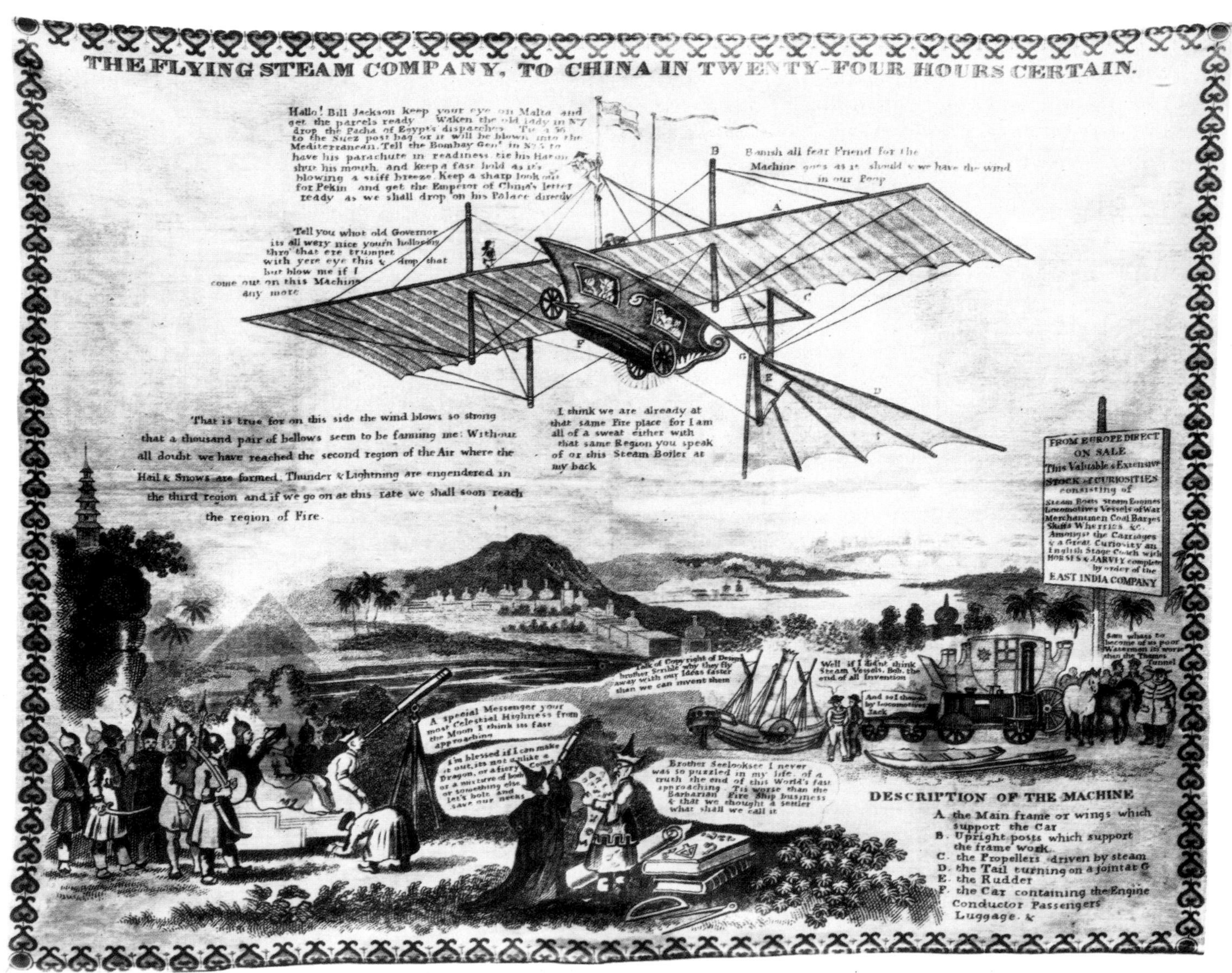

THE FLYING STEAM COMPANY, TO CHINA IN TWENTY-FOUR HOURS CERTAIN.
Hallo! Bill Jackson keep your eye on Malta and get the parcels ready Waken the old lady in Nº7 drop the Pacha of Egypt's dispatches ... to the Suez post bag or it will be blown into the Mediterranean. Tell the Bombay Genl in ... to have his parachute in readiness tie his ... shut his mouth and keep a fast hold as it's blowing a stiff breeze. Keep a sharp look out for Pekin and get the Emperor of China's letter ready as we shall drop on his Palace directly
Banish all fear Friend for the Machine goes as it should & we have the wind in our Poop
Tell you whot old Governor its all wery nice youn hollaring thro' that ere trumpet with yere eye this ... drop that but blow me if I come out on this Machine any more
That is true for on this side the wind blows so strong that a thousand pair of bellows seem to be fanning me: Without all doubt we have reached the second region of the Air where the Hail & Snows are formed. Thunder & Lightning are engendered in the third region and if we go on at this rate we shall soon reach the region of Fire.
I think we are already at that same Fire place for I am all of a sweat either with that same Region you speak of or this Steam Boiler at my back
FROM EUROPE DIRECT ON SALE This Valuable & Extensive STOCK of CURIOSITIES consisting of Steam Boats Steam Engines Locomotives Vessels of War Merchantmen Coal Barges Skiffs Wherries &c. Amongst the Carriages is a Great Curiosity an English Stage Coach with HORSES & JARVEY complete by order of the EAST INDIA COMPANY
Talk of Copyright of Design brother Scribble why they fly away with our Ideas faster than we can invent them
Well if I didn't think Steam Vessels, Bob, the end of all Invention
And so I thought by Locomotive Jack
A special Messenger your most Celestial Highness from the Moon I think its fast approaching
I'm blessed if I can make it out, its not unlike a Dragon, or a fiery Comet or a mixture of both or something else let's bolt and save our necks
Brother Seelooksee I never was so puzzled in my life, of a truth the end of this World's fast approaching 'Tis worse than the Barbarian Fire Ship business & that we thought a settler what shall we call it
DESCRIPTION OF THE MACHINE
A. the Main frame or wings which support the Car
B. Upright posts which support the frame work.
C. the Propellers driven by steam
D. the Tail turning on a joint at G
E. the Rudder
F. the Car containing the Engine Conductor Passengers Luggage. &c

2

3

1. Henson's model of his "Aerial Steam Carriage." 2. Souvenir pillow cover showing Henson's *Ariel*. 3. William Samuel Henson. He designed his steam-powered flying machine in 1841, but built only a model of it, which failed to fly.

naphtha and its crew of three aboard, it weighed some eight thousand pounds.

Maxim first made some tentative trials with his aerial pantechnicon and found that even at part throttle it tended to lift off the rails—to fly, in fact. To prevent this he built wooden tracks about two feet above the rails and fitted extra wheels on outriggers from the machine's frame. These engaged the wooden track from below and prevented takeoff.

In July, 1894, Maxim essayed a full-power test. The monster tried to fly. As it started to rise, it broke the restraining tracks, which became entangled in the struts and wires. Maxim signaled for the steam to be shut off and the great flying machine settled to the ground only minimally damaged. Smugly satisfied that he had proved that a machine could fly, Maxim said, "Propulsion and lifting are solved problems; the rest is a mere matter of time." Like others after him he assumed that a successful aircraft would be inherently stable. In his book, *Artificial and Natural Flight,* he says: "In regard to the stability of the machine, the centre of weight is much below the centre of lifting effect; moreover, the upper wings are set at such an angle that whenever the machine tilts to the right or to the left the lifting effect is increased on the lower side and diminished on the higher side. This simple arrangement makes it automatic as far as rolling is concerned." Maxim was also of the opinion that "it will be necessary to steer in a vertical direction by means of an automatic steering gear controlled by a gyroscope." Luckily, Sir Hiram never embarked in a flying machine dependent on such "automatic" controls.

The idea that once you managed to lift a flying machine off the ground it would just fly along on an even keel died hard. Man first had to learn something about managing wings in the turbulent ocean of air above him before he could successfully apply power to them. One of the first to realize this was the great Otto Lilienthal. Lilienthal had been imbued with the idea of flight since boyhood, when he watched the storks soaring above the chimney tops of his home in Pomerania. By 1891 he was building his first glider. If he could learn to fly, he would worry about a means of mechanical propulsion later.

Unlike modern glider pilots, Lilienthal did not sit inside his plane. He hung from a single cambered wing, his head and shoulders above it. His controls were his own muscles. By swinging and twisting his body he was able to influence the attitudes of his gliders in the unpredictable air. He had no ailerons and his tail planes were not movable. His feet and legs were his landing gear.

He had to take off from a height into the wind, of course, and as natural hills do not always slope in the direction of the prevailing breeze, he built himself a cone-shaped hill from which he could always face the wind. Between 1891 and 1896 he took to the air more than two thousand times, often achieving glides of several hundred feet. In 1896 he thought the time had come to try powered flight. He built a glider which was part ornithopter. Its wing tips, powered by a small engine worked by compressed carbonic-acid gas, were to flap like a bird's. But before this strange craft was ever proved, Lilienthal, flying one of his normal gliders in rough air, stalled, crashed, broke his back, and died.

Lilienthal's exploits not only brought him worldwide fame, but inspired men in many countries to emulate him. For Lilienthal had demonstrated that brave, winged men could at last ride the air. Others—the Scotsman Percy Pilcher, who also died flying a glider in 1899, and Octave Chanute, who was too old and no longer had the reflexes and strength required for gliding, but who designed and built gliders and who published the important book, *Progress in Flying Machines,* in 1894—also carried on Lilienthal's kind of flying experiments. These men were concerned more with control in the air than with engines, air frames, and inherent stability—problems which still engaged earthbound theoreticians sweating over their drawing boards.

In America, meanwhile, Samuel Pierpont Langley, a portly, elderly, bearded gentleman, a famous astronomer and director of the Smithsonian Institution, had built a flying machine which still stirs controversy.

Langley, much influenced by Penaud's work, first experimented with rubber-band-motored model "Aerodromes" (his odd name for aircraft) in 1887. But they were capable only of erratic flights of a few seconds' duration. In

1

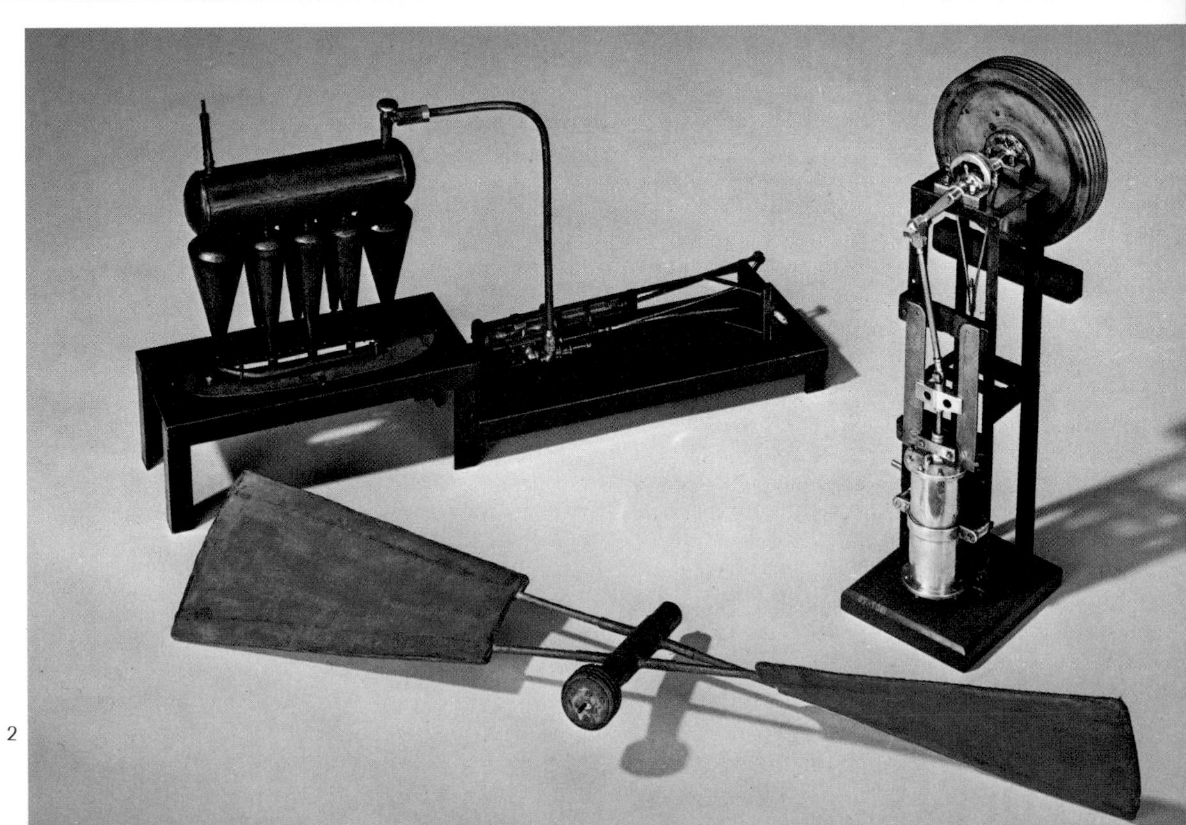

2

1. John Stringfellow's model triplane of 1868 had great influence on aeroplane design.
2. Stringfellow's aeronautical steam engines, boiler, and propeller of 1843-48.
3. Alphonse Penaud's 1876 design for an amphibious aeroplane with retractable wheels.
4. Model of Clément Ader's steam-powered *Eole*. The covering of one wing is removed to show the structure.

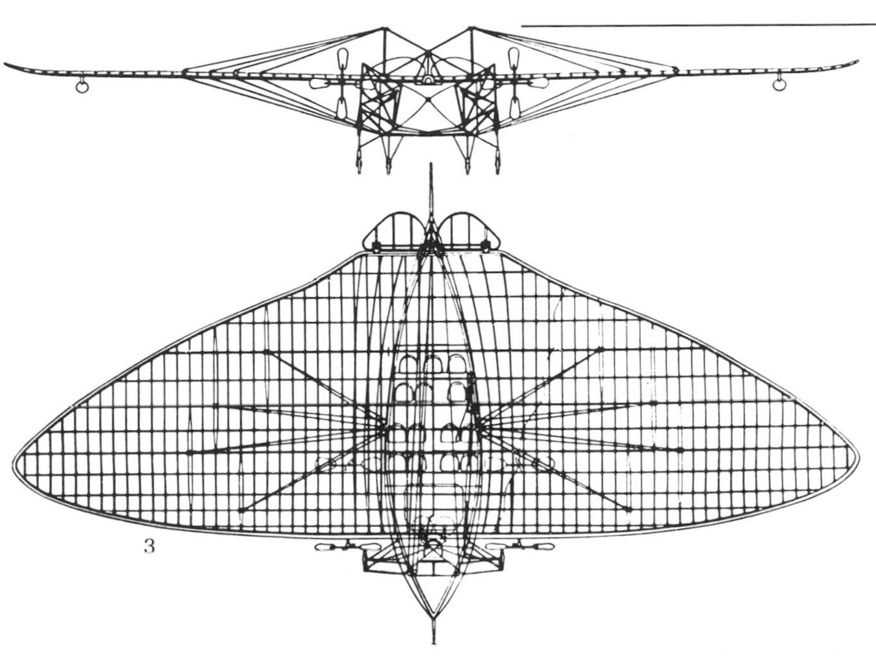

3

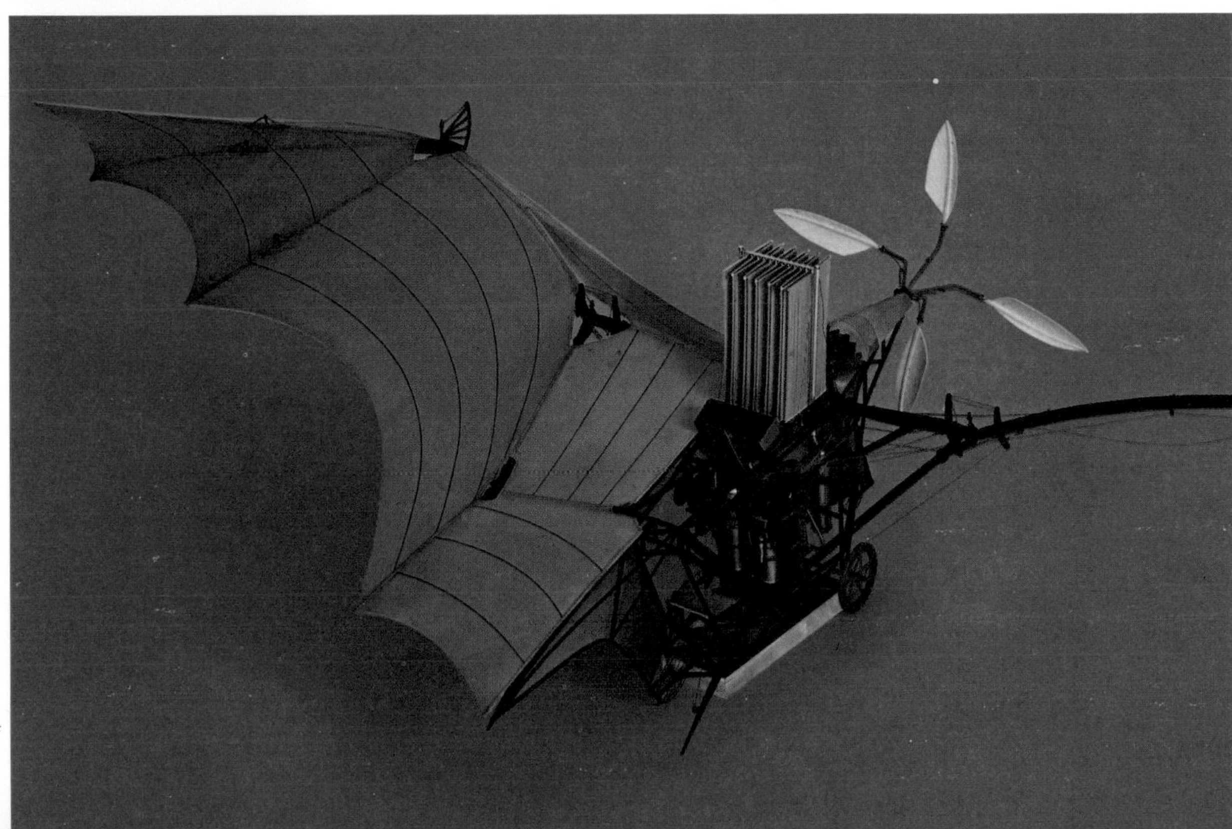

4

1. Sir Hiram Maxim.
2. Maxim's model of his flying machine (rear view).
3. Complete with top hats, watch chains, and umbrellas, members of the Royal Aeronautical Society come to view Maxim's giant machine. 4. The machine on its track. 5. Maxim's compound steam engine.

1

2

4

3

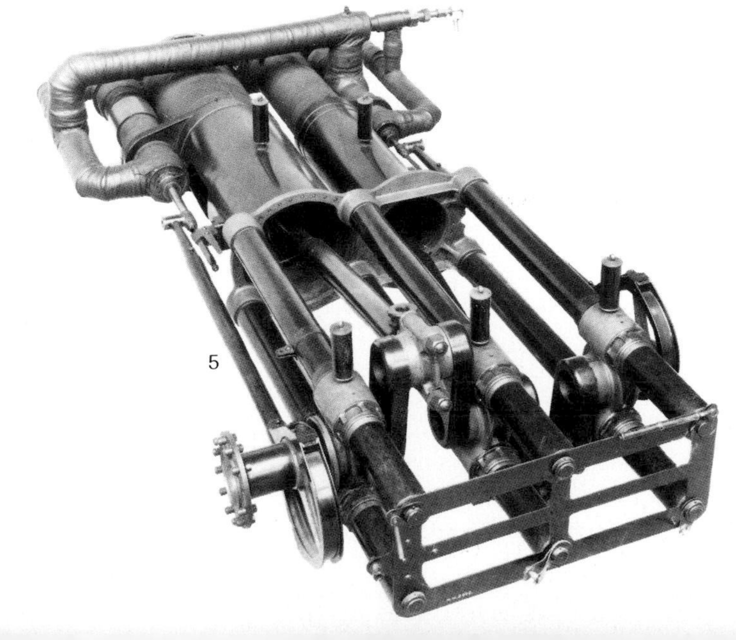
5

Otto Lilienthal, the great German gliding pioneer, showed that man had first to learn how to manage wings in the air before he dared apply engine power to those wings. All who came after him, including the Wright brothers, benefited from his daring experiments. He crashed and died in 1896.

1891 he turned to steam as his motive power. He spent five years designing, redesigning, building, and rebuilding a succession of model Aerodromes, boilers, and engines, with the help of his devoted and brilliant assistant, Charles M. Manly, before achieving any success. Finally in May, 1896, in the presence of Langley's friend Alexander Graham Bell, one of the Aerodromes launched from a houseboat in the Potomac River near Quantico, Virginia, flew 3,300 feet while Mr. Bell photographed it in the air. In November, a flight of 4,200 feet was achieved. These complex models of Langley's, with single wings set one behind the other and their engines and boilers amidships, were by no means toys (they had a six-foot wingspan), but it was fashionable then to deride flying-machine inventors and Langley was laughed at for wasting time and money on silly gadgetry.

Later, Manly wrote of the flights: "They meant success! For the first time in the history of the world a device produced by man had actually flown through the air, and had preserved its equilibrium without the aid of a guiding human intelligence."

In 1898 the United States and Spain went to war. President McKinley, aware of Langley's success with models, had the odd hope that Langley could somehow whip out a flying machine for fighting against the backward Spaniards. He prodded the War Department into advancing $50,000 to build it.

Langley gave up the idea of using steam power and decided that a 12-hp gasoline engine weighing under a hundred pounds would spin the propellers of his full-size machine.

It was poor Manly's job to have such an engine built, although the gasoline engines available for the primitive motor cars of the time were too feeble, too crude, and too heavy to meet Langley's power and weight needs.

After a long, painful search, it was decided to give Stephen M. Balzer in New York the contract to build the engine: a five-cylinder radial engine with rotating cylinders, a design far in advance of its time. Balzer just about bankrupted himself doing it. A smaller engine for a quarter-size model was also to be built. Incredibly, Balzer agreed to finish the larger engine by February, 1899. He actually had it almost finished in time, but, sadly, he couldn't make it run very well. In May, 1900, after Balzer had spent a year and a half frantically making changes, Manly visited his shop and found that it developed only 2.83 horsepower and then only for a few minutes, after which it inexplicably died.

Manly decided to build the engine himself in the Smithsonian's shop using some of Balzer's parts. After considerable redesigning of the ignition system, the water jackets, and a hundred other elements, he succeeded in extracting 50 hp at some 825 rpm from the 120-pound engine by the time it was installed in the full-size Aerodrome, an almost incredible power-weight ratio for its day.

Manly's peculiar carburetor is worthy of note. It consisted of a small tank filled with lumps of porous wood saturated with gasoline from which the engine sucked the vapor.

While the man-carrying Aerodrome was under construction, it was decided to fly the gas-engined quarter-scale model. On August 8, 1903, this model, the first flying machine in history to be propelled by a gasoline engine, was successfully flown off a houseboat in the Potomac River. Although it flew only a thousand feet, Langley and Manly were jubilant, especially over the inherent stability of the model.

Manly wrote: ". . . in order to traverse at will the great aerial highway man no longer needs to wrest from nature some strange mysterious secret, but only, by diligent practice with machines of this very type, to acquire an expertness in the management of the aerodrome not different in kind from that acquired by every expert bicyclist in the control of his bicycle." How wrong poor Manly was!

In October the full-size Aerodrome, a sort of double monoplane, was at last ready for its first flight. It had been manhandled piecemeal atop a complex spring-operated launching catapult which was mounted upon the roof of a big houseboat in the Potomac. Thin, scholarly little Manly started the engine, which roared healthily through its open exhaust, and bravely climbed aboard the 750-pound machine. (Langley was in Washington on other business.)

The *Washington Post* next day carried its version of

Otto Lilienthal.

1

2

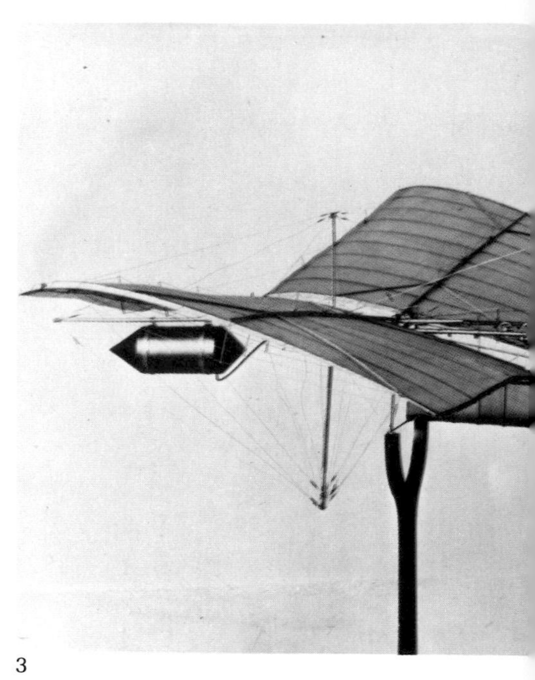
3

4

5

6

7

1. Samuel Pierpont Langley.
2. Manly-Balzer engine.
3. Model of Langley's Aerodrome.
4. Aerodrome poised for takeoff from a barge on the Potomac.
5. Charles M. Manly with Langley before a flight.
6. A six-foot model of the Aerodrome flew 3,300 feet in May, 1896.
7. The full-size Aerodrome crashed on takeoff and plunged into the Potomac.

the debacle: "Manly looked down and smiled. Then his face hardened as he braced himself for the flight, which might have in store for him fame or death. The propeller wheels, a foot from his head, whirred around him one thousand times to the minute. A man forward fired two skyrockets. . . . A mechanic stooped, cut the cable holding the catapult; there was a roaring, grinding noise—and the Langley airship tumbled over the edge of the houseboat and disappeared in the river . . . it simply slid into the water like a handful of mortar." The sodden Manly was plucked from the Potomac.

The Aerodrome had somehow been hung up on Langley's crazily complicated launching mechanism. But Langley dearly loved his catapult and in December another attempt was made. This time the aircraft's tail surfaces became entangled in the launching mechanism and were torn off. The machine first tried to climb vertically, then fell over on its back and plunged into the river. Happily, Manly was again fished out, but the Army withdrew further support. Langley's wonderful Aerodrome went into storage at the Smithsonian Institution.

Would the Aerodrome finally have flown? Perhaps. But not if Langley had continued to pursue the chimera of inherent stability.

Nine days after Langley's Aerodrome fell into the Potomac, two spare, quiet brothers from Ohio flew the world's first successful airplane.

The Wright brothers had followed a different course from Ader or Maxim or Langley. They were fliers more than mechanics and theoreticians. They were the descendants of Lilienthal and Chanute. They learned to ride the deceitful air like the hawk or the albatross before trying to push their way through it with a stinking internal-combustion engine.

Unlike Langley, they had no great national institution behind them. The U.S. Army spurned them until many years later, when they were already flying figure eights over the brass hats.

Orville and Wilbur Wright were small boys when Alphonse Penaud was still engrossed with problems of flight. He had invented a rubber-band-operated helicopter—a toy —a tiny device of bamboo and tissue paper. In 1878, the Wright boys' father, a clergyman, brought one home to the children, who flew it until it disintegrated. Wilbur, the elder, is said to have tried to build improved versions of Penaud's model and to have discovered that the bigger he made them the worse they flew—not realizing that a machine only twice the size required eight times as much power to fly. This may be an apocryphal story, like that of James Watt and the teakettle, but as they grew older the Wright brothers showed a more than usual interest in things mechanical.

By 1892, when Wilbur was twenty-five and Orville twenty-one, they were involved with the bicycle craze that was sweeping the country. They opened a shop in Dayton, Ohio, where they sold and repaired bicycles, and a few years later they started manufacturing them.

Perhaps it was their childhood experience with the Penaud helicopter that started them thinking about flight. In any event the brothers eagerly started reading everything they could find on the subject, although there wasn't much in Dayton, Ohio, in the 1890s. But Otto Lilienthal's gliding exploits and his fatal crash in 1896 created such a sensation that news filtered even into the fastnesses of Middle America. The Wrights devoured everything they could find to read about Lilienthal and everything available on gliding and flying. Soon they learned of the experiments and calculations of Octave Chanute, Langley, Maxim, Ader, and Penaud. (They also found that almost everything written about flying was wrong.)

In 1899 the Wrights made their great discovery. They observed that soaring birds—there were buzzards to watch near Dayton—"regain their balance . . . by a torsion of the tips of their wings." They decided to apply this bird movement to their first glider's wings. They considered pivoting the wings, but realized that such a mechanical solution would be structurally weak.

They pondered methods of wing control, until one day Wilbur sold a bicycle-tire inner tube. He took the tube out of its cardboard box and while talking to the customer, idly twisted the long box. He immediately saw the solution to their problem. The wings of their glider could be twisted—"warped." As Orville said,". . . the basic idea was the adjust-

1. 1901 Wright glider, flown as a kite.
2. 1902 Wright glider at Kitty Hawk.
3. Wilbur and Orville Wright.
4. The modified 1902 glider soars from Kill Devil Hill on October 21, 1903. Note the Wrights' camp in the distance.

1

3

4

2

1. The Wrights' 12-hp, four-cylinder engine weighed 179 pounds.
2. The end of Wilbur Wright's first attempt at flight, December 14, 1903. He overcorrected with the elevator, and the *Flyer* plowed into the sand. 3. The world's first powered, sustained, and controlled flight. Orville Wright is the first man to fly, on December 17, 1903.

1

3

ment of the wings to the right and left sides to different angles so as to secure different lifts on the opposite wings." This was the decisive invention upon which the Wrights' patents were based and from which all later methods of lateral control by ailerons sprang.

When they built their first model glider in 1899—it was really a sort of kite with a five-foot wingspan—it was a biplane with struts and bracing similar to those built by Chanute—the only bit of technology, incidentally, that they borrowed from him.

Watched by some small boys, they flew their kite in an open field just outside Dayton. By means of cords controlled by sticks in their hands, they could warp the wings from the ground and also move the upper wing to control the center of lift. The kite also had a sort of elevator at its rear.

The next step was a glider that could carry a man. The brothers now needed a better place than a Dayton meadow for their experiments—a treeless place with hills they could glide from and with steady winds to help them stay aloft. They wrote the U.S. Weather Bureau which sent them a list of several such desolate and windy places. They chose Kitty Hawk, North Carolina.

The first man-carrying glider was finished by September, 1900. It had taken the Wrights only a few weeks to build after the summer bicycle business slacked off and had cost under $15, plus the brothers' labor. By October they had it flying at Kitty Hawk, mostly as an unmanned kite, although a few free glides were also attempted. But this glider didn't fly as well as they had hoped, requiring much more wind than their calculations (still based on those of Lilienthal) had indicated. But they had proved to themselves that their method of wing warping worked, although their glider had flown as a kite with a Wright lying prone on the lower wing for only ten minutes and as a free-manned glider for a total of some two minutes.

They decided to build a bigger glider for the 1901 season, abandoning the glider they had been using and giving the sateen wing covering to a local lady, who made it into dresses for her two little girls. Neighbors thought it rather a shame that they had used such nice material on a mere kite.

The 1901 glider, flown from Kill Devil Hill, four miles from Kitty Hawk, proved to be cranky in the air, in spite of its larger wing area and greater wing curvature. Old Octave Chanute journeyed out to the bleak North Carolina sandspit to encourage the Wrights, but still their spirits flagged. "Nobody will fly for a thousand years!" grumped Wilbur.

In this glider the operator lay in a cradle to which were attached the wires that warped the wing tips. By twisting his hips he controlled the warping.

But the problem was wing shape. The Wrights found that they still did not know enough about aerodynamics, that nearly all of the existing data was incorrect. Later they recalled, "We were driven to doubt one thing after another, till finally, after two years of experiment, we cast it all aside."

During the next winter at Dayton they started at the beginning. They built themselves a little wind tunnel sixteen inches square and in it they tested a series of miniature wings.

Now they knew how an airplane's wing really ought to be shaped; flatter and narrower than anyone before them had assumed to be correct. Still another glider resulted from their experiments. And it flew the way they had hoped it would fly. In October-November, 1902, the Wrights made almost a thousand glides, one of them over a record distance of 622½ feet and a duration of twenty-six seconds.

At first they had trouble. The glider had a forward elevator and a pair of fixed fins aft, but no movable rudder. Warping the wings to bank in order to turn often did not work. The raised wing would hang back and as the lowered wing shot ahead, the glider swooped down into the sand. Quickly reversing the warping to correct this caused spins.

Orville came up with the idea of replacing the fixed rear fins with a movable rudder. Wilbur then suggested that the new rudder controls be connected to the hip-operated cradle which warped the wings. Now air pressure against the rudder, which was always turned toward the warped wing, automatically counteracted the drag. With this technique the Wrights solved their problem of control in the air. They were now convinced, too, that their flying machines should not be inherently stable, but should rather be sensitive and quickly

Below: After successful flights at Kitty Hawk in 1903, the Wrights continued their flying experiments at Huffman Prairie, near Dayton, Ohio, in 1905.
Right: Diagrammatic drawings of the first Wright *Flyer*. Note the reverse dihedral of the wings and the off-center mounting of the engine, which allows the pilot to lie beside it.

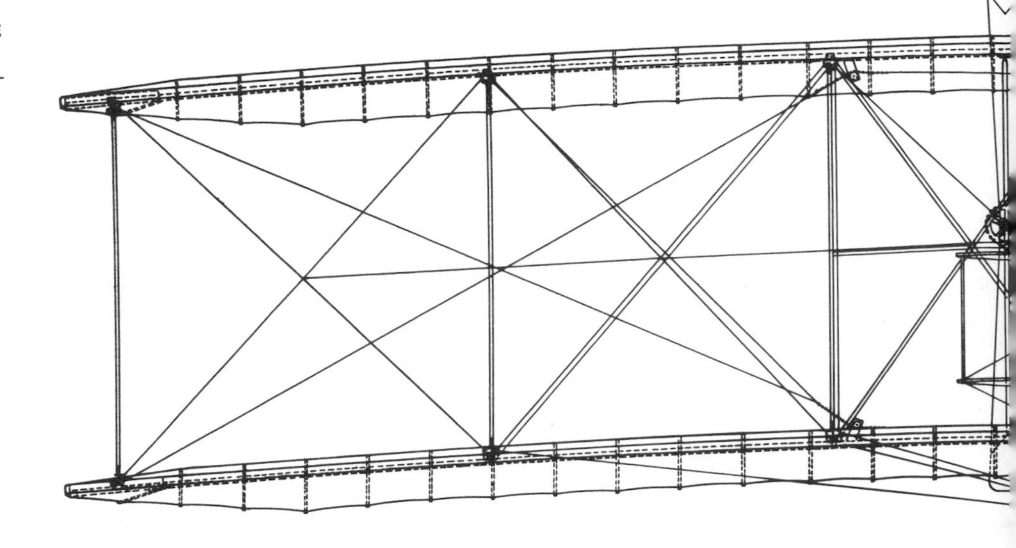

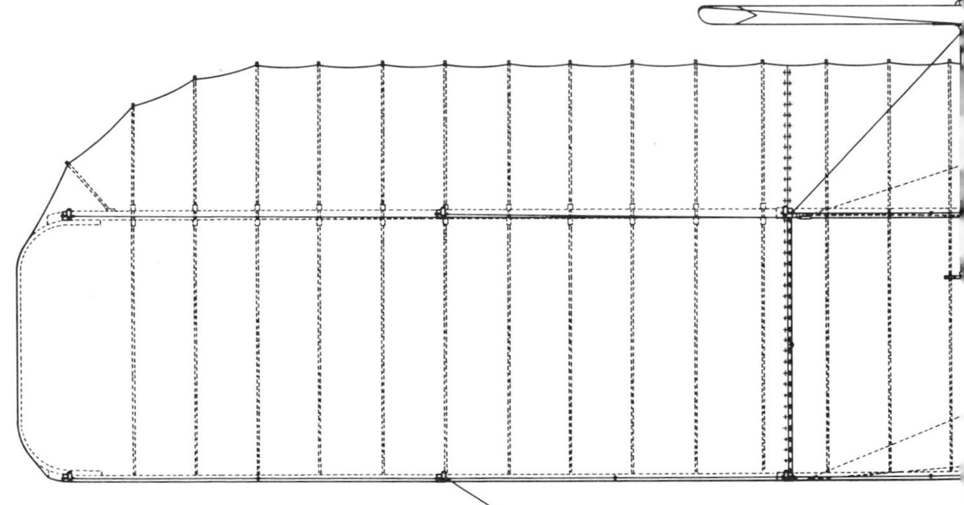

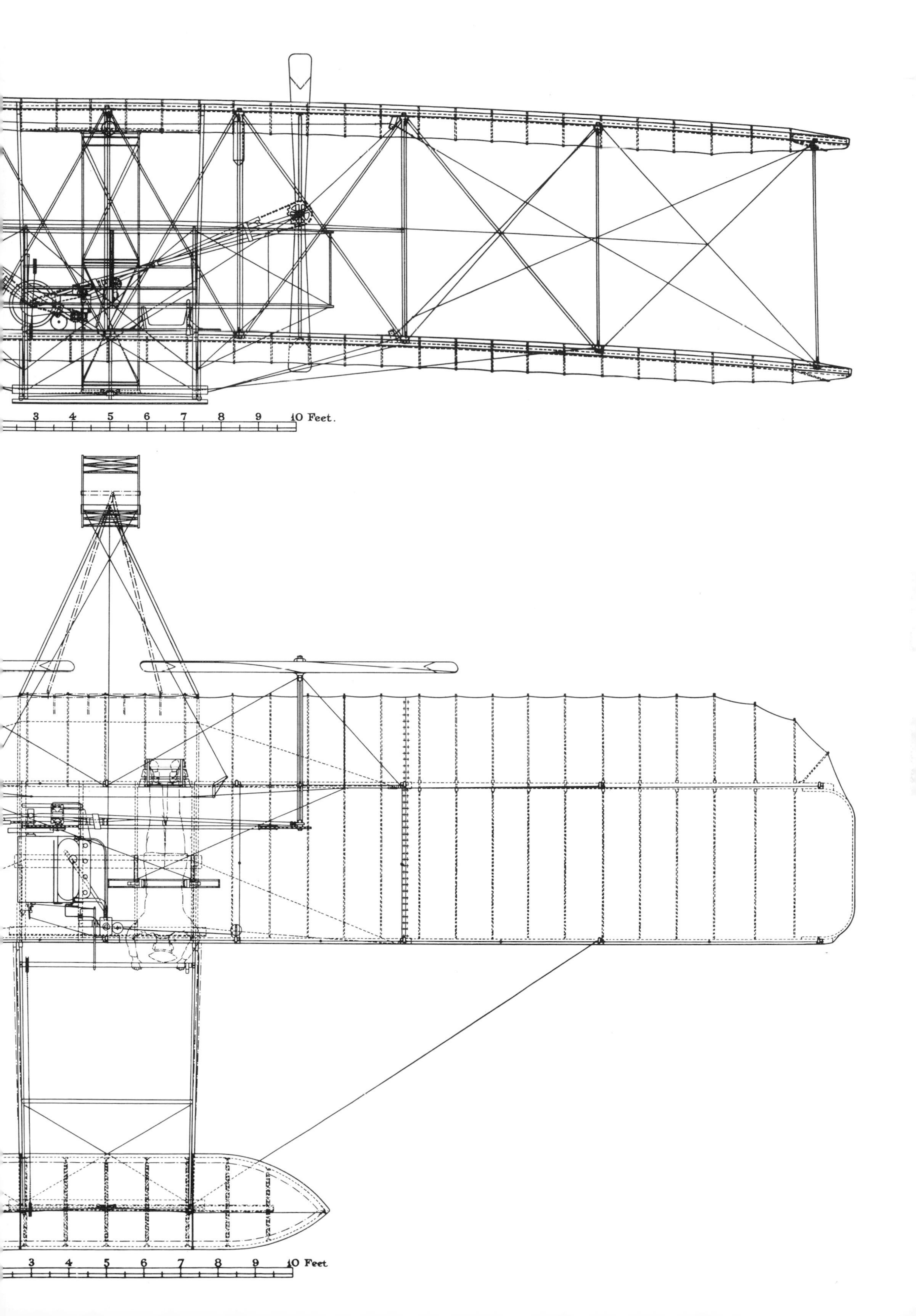
3 4 5 6 7 8 9 10 Feet.
3 4 5 6 7 8 9 10 Feet

The French did not entirely believe that the Wrights had been flying successfully while they themselves had only been able to make feeble hops. In 1908 Wilbur went to France and proved the superiority of the Wright aeroplanes.
Right: He explains his machine to Edward VII of Britain.
Opposite: Wilbur demonstrates with a passenger at Pau, France, in 1909. King Alfonso of Spain watched.

responsive to a pilot's wishes.

Now the Wrights could climb or descend by means of their front elevator, and with rudder plus wing-warping they could maintain their horizontal balance and also make smooth turns.

What they needed next was an engine and propellers to pull their machine through the air fast enough to keep it aloft. They wrote to several early automobile builders asking if they could supply an eight-horsepower engine weighing under two hundred pounds. Most replied that they were too busy building engines for their cars. Some seem to have suspected the Wrights of wanting to use the engine for a flying machine and dared not have their names coupled with such a crazy project. In the end the Wright brothers built their own engine with the help of their mechanic, Charlie Taylor. It had four cylinders of four-inch bore and stroke. Water-cooled, it had five main bearings. Ignition was by low-tension magneto. It developed 12 hp at 1090 rpm but after a few minutes power dropped to 9 hp due to the inlet air being heated by the water jacket. It weighed 179 pounds.

Next they had to design propellers. They had tried to base their experiments on what was known about the characteristics of ships' propellers, but discovered that the design of marine screws was pretty much "by guess and by God." Using their wind tunnel they at last arrived at shapes for their air screws which were much more efficient than those of Maxim or Langley.

The little engine drove the twin propellers by means of long bicycle chains, one of which was crossed to make the propellers spin in opposite directions. The chains ran inside tubes.

The Wrights did not merely hang this engine and propellers on their 1902 glider. They built an entirely new machine for their powered flights. It was, like the gliders, a biplane. It had a 40-foot, 4-inch wingspan, a wing area of 510 square feet, a biplane elevator out front, and a double rudder aft. It weighed 605 pounds empty. And it had a name by which all later Wright aeroplanes were also called: the *Flyer*.

The brothers and their *Flyer* arrived at Kitty Hawk in late September, 1903. It took them three weeks to uncrate and assemble the machine, during which time they got out their 1902 glider and took it up for a few practice flights.

The new *Flyer* had teething troubles. The chain sprockets on the propeller shafts kept coming loose. Being bicycle men they had the obvious solution. They stuck the sprockets on firmly with tire cement. A tubular propeller shaft cracked when the engine backfired on test and Orville went back to Dayton to have a new pair of shafts machined. Then the weather turned foul. It wasn't until December was half gone that the *Flyer* was ready for action.

The *Flyer* did not have a wheeled landing gear. It was, like the gliders, fitted with skids for landing on the sand. To take off it was set on a crossbeam fitted to a little wooden trolley which ran on a portable wooden rail. A wire held the machine in place as the engine was run up to its peak revs. Then the wire was let go and the *Flyer* was supposed to zip down the track and into the air, leaving the trolley on the ground.

The *Flyer* was not catapulted into the air as some Frenchmen later claimed. After three-quarters of a century, one of them, the quixotic automobile manufacturer Gabriel Voisin, just before he died at ninety-three, still maintained that his aeroplane, which made a six-second flight in 1907, was the first to make an unassisted takeoff.

On December 14, after waiting for two days for a good breeze, Wilbur, who had won the toss of a coin, attempted a flight. The track had been laid on the slope of a dune to gain momentum. A few hardy residents of Kitty Hawk stood watching and shivering in the wind as Wilbur lay down on the wing and revved the engine. The *Flyer* rushed down the forty-foot track. But Wilbur tried to climb too soon, stalled, and after 3½ seconds in the air came down in the sand. A skid and a few other minor parts were broken.

On December 17, the damage made good, the brothers lay down their track again facing into a 21-mph wind. The track was on level ground this time. They also alerted the men at a life-saving station down the beach, and the crew of five came to manhandle the unwieldy *Flyer* out of its hangar and onto the track. The wind by now was strong enough to blow the dune-grass flat and raise stinging sand off

the dune tops. It was Orville's turn, and he settled himself prone into the control cradle on the wing beside the engine. The propellers became circular blurs and the flying machine strained against the wire. Orville released the wire's catch and the *Flyer* on its trolley started to move down the track into the wind as Wilbur ran to keep up with it. As it reached the end of the track Orville cautiously operated the elevator. Now he was flying. He flew for twelve seconds into the hard wind. One hundred and twenty feet down the beach he landed safely in the sand. Man had at last found his wings.

The Wrights took to the air three more times that morning. Wilbur, taking the *Flyer* up for the last flight of the day, stayed aloft for fifty-nine seconds and flew a distance of 852 feet. Since it was breasting half a gale, it traversed three times that distance through the fast-moving air into which its propellers labored to push it. The Wrights knew exactly how much air the *Flyer* had moved through, for they had foresightedly carried along an anemometer built by Jules Richard of Paris.

It was several years before anyone did more than hop off the ground in a flying machine. Santos-Dumont managed to hop-fly for 21 1/5 seconds in 1906. In 1907 the Frenchman Henri Farman, in a Voisin-Farman, stayed in the air for one minute and fourteen seconds. By then the Wrights were making flights of more than half an hour from a cow pasture near Dayton, Ohio.

After the fourth flight, after the wind unhappily caught the *Flyer* on the ground and rolled it over and over, smashing it so that no more flights were possible that year, Orville sent a telegram to his father. With the telegraph company's expected mistakes, it read: SUCCESS FOUR FLIGHTS THURSDAY MORNING ALL AGAINST TWENTY ONE MILE WIND STARTED FROM LEVEL WITH ENGINE POWER ALONE AVERAGE SPEED THROUGH AIR THIRTY ONE MILES LONGEST 57 SECONDS INFORM PRESS HOME CHRISTMAS

OREVELLE WRIGHT

6

6 Light Does the Drawing

Pictures for the Multitude

Life in the contemporary western world would be inconceivable without photography. The things we eat, drink, and wear, our pleasures and accouterments, the houses we live in and *their* furnishings and appliances—all are exhibited and sold to us largely by photography. Through images cast by lenses on bits of chemically coated celluloid, entertainers entertain, politicians curry favor, news and information are transmitted, attitudes are struck, and social and cultural patterns are established.

Astronomers photograph pinpoints of light which began their journey through space before the birth of earth. Biologists using electron microscopes photograph the infinitely small chromosomes of life and the microbes of death. Physicists photograph the essential particles of matter. Satellites rocketed into orbit click their cameras both to spy upon the world and to search out the secrets of its distant sister planets. And billion-dollar industries thrive on the production of moving pictures.

Early photographic experimenters and inventors had no inkling that their opticochemical means of image-making would become so powerful a force. To them photography was merely a new way to make pictures, a mechanical method by which those untutored or untalented in drawing or painting could equal, or at least approximate, the work of professional artists.

Before photography, people were dependent on the artist for pictorial representations of reality—for visions of other people, places, things. They were dependent not only on the degree of each artist's talents, but on the breadth of his experience and the taste of his patrons, as well. Those who relied on Albrecht Dürer for their notion of a rhinoceros got beautiful draftsmanship but outlandish zoology.

In the western world, public art—the art accessible to common folk—was generally underwritten by the church or by king and court. Private art was a luxury, affordable only by the wealthy and consisting largely of portraiture. There were other genres, but most art was highly provincial. Most artists were not widely traveled and were ignorant of, or grossly misinformed about, exotic people and places.

In 1839, when the daguerreotype process was revealed to the world, Victoria had been queen a mere two years, Martin Van Buren was president of twenty-six states, Dickens had just written *Nicholas Nickleby,* and surgery was practiced without anesthesia. The parachute, friction match, digital calculator, and Braille had been invented, but the cylinder lock, linoleum, condensed milk, and the safety pin had not. In the United States that year Charles Goodyear first vulcanized rubber.

Photography's impact on this world was instantaneous and tremendous. Almost immediately, publishers sent photographers heavily laden with primitive equipment to point their lenses at reality: the ruins of Egypt, the Great Falls of the Niagara, the Mexican War, celebrities like British monarchs and American presidents, even illicitly posed burlesque queens in tights. Yet almost to the beginning of the twentieth century, such pictures could be seen only in the form of original photographic prints, or as wood engravings gouged out of hard maple blocks.

Not until the photoengraving process was developed did it become possible to reproduce pictures by the millions in newspapers and magazines.

However great its power for information, education, and propaganda, photography since the days of calotypy and daguerreotypy has also been an art form, a means of graphic expression. Since 1839 pundits have pointlessly argued the question, "Is photography an art?" And today prints by big-name photographers are collected as avidly as Rembrandt drawings.

Still, most photographs are made by plain people, not for advertising or publishing Pooh Bahs, or for political propagandists, or even by scientists peering into oculars. Nor do they have artistic pretensions. And what do people photograph after more than a century of photographic invention and research?

Mostly babies and pussycats.

Preceding pages: Most American daguerreotypes were portraits. Made by the million, they could be had for as little as twenty-five cents. Opposite: 1630 camera obscura, used for examining sunspots. Left: Nineteenth-century tent-type camera obscura as used by a landscape painter.

The Silver Image

It is a not-unreasonable guess that about the time mankind began to regard itself approvingly in the first mirrors of highly polished Bronze Age bronze, it also yearned for a means to retain the dear reflected image permanently. Certain it is that by the sixteenth century, when the camera obscura was first contrived, artists tracing the images its lenses threw onto its ground-glass screen wondered if there were not a way to transfer these views—again permanently—to paper.

These hopes eventually were realized and the wish made real in the 1830s, when, to universal jubilation, daguerreotype images were fixed on a mirror-like silver plate and calotypes on paper. There was carping as well: Daguerreotypes and calotypes lacked the full colors of nature, their processes were too slow, they required too many calculations and too much messing about with smelly chemicals.

Now, a century and a half later, we have overcome even these objections with Polaroid's SX-70, undoubtedly only the first of many cameras designed to deliver a dry full-color picture every few seconds. And, moreover, there is as great a contrast between the ways Louis Jacques Mandé Daguerre and Edwin H. Land invented their photographic processes as there is between the processes themselves.

Although daguerreotypes and calotypes were the first *practical* photographs, they were not the first photographs. It had long been known that silver salts could be darkened by exposure to light. Johann Schulze of the University of Altdorf in Germany had demonstrated this as early as 1725. But it was not until some seventy-five years later that Thomas Wedgwood, of the famous family of potters, tried to capture the images formed by the camera obscura on paper moistened with nitrate of silver. He failed. Although he gave his sensitized paper long exposures, his images were unusably faint. And even these turned black when exposed to light outside the camera. Next he tried sensitizing white-painted leather and placing "the woody fibres of leaves, the wings of insects" in contact with the leather and exposing the superpositions to bright sunlight. Here he got contact prints, but these, too, soon blackened all over. The great Sir Humphrey Davy desultorily collaborated with Wedgwood, but he also failed to come up with a fixing agent to render the images permanent.

Fifteen years later, soon after Waterloo, a man across the English Channel had somewhat better luck. This was Joseph Nicéphore Niepce of Chalon-sur-Saône in France.

It was lithography which first led him and his son Isidore to investigate photographic problems. This is the process, invented by Alois Senefelder in Munich in 1798, whereby drawings in greasy ink or crayon on smooth blocks of Solenhofen limestone can be beautifully transferred to paper. By 1813 lithography had become a veritable craze in France, and Niepce, a born experimenter, was inevitably engrossed by the new art. But, since Solenhofen stone was hard to come by in France, Niepce used pewter plates instead. Nor was he much of an artist. He lacked the ability to draw on the plates. He tried to get the sun to do the drawing by letting it shine through waxed, translucent engravings laid on the plates, which he tried to sensitize with various varnishes he hoped would retain an image—at first without much luck.

By 1816 Niepce seems to have given up on his lithographic experiments. Now he tried to make his pictures in a camera, using paper sensitized with silver chloride. He wrote his brother Claude: "I've had made a kind of artificial eye that is quite simply a little square box six inches on each side, fitted with an extension tube that carries a lens." But he dropped and broke this first of all photographic cameras and had to make another. He wrote again: "I put the apparatus in the room where I work in front of the window opposite the birdhouse. I made the experiment with the process that you know, *cher ami*, and I saw on the white paper all the parts of the birdhouse." He went on to complain that the image was reversed in tone—the light areas in the scene were dark, the dark ones light. The simple idea of contact-printing the negative image on another piece of paper to reverse the tones had not occurred to him. Anyhow, he had not found a means

1. 1826-27 photo-etched plate of Cardinal d'Amboise by Nicéphore Niepce.
2. 1826 heliograph on pewter by Niepce.
3. Portrait of Nicéphore Niepce.
4. World's first photograph, taken from Niepce's window at Gras in 1826.
5. Photograph on glass of a table setting, taken by Niepce about 1829.

1

2

of fixing an image.

Niepce thought that if he could find some chemical that light would turn white instead of black, he would have at least one of his problems licked. After much search he settled on a substance called bitumen of Judea, which normally was soluble in lavender oil, but became insoluble if exposed to light. On a plate coated with bitumen, the parts of a picture lightest in tone would remain, the darkest parts would be entirely washed away by the lavender oil, and the middle tones would be partly dissolved.

Niepce's continued experiments with the Pyréolophore engine seem to have slowed down his work in photography. It was not until 1822 that he at last had a remarkable success. He coated a sheet of glass with bitumen of Judea, placed a waxed engraving of Pope Pius VII upon it, gave it a lengthy exposure to sunlight, and achieved a photographic copy of the engraving on his glass plate—a heliograph, as he called it.

Later he learned to contact-print engravings on pewter. By acid-etching the dark parts (those washed clear of the bitumen by oil of lavender), he succeeded in making printing plates. The most successful of these is a copy of an engraving of Cardinal d'Amboise, a minister of Louis XII's. Prints from it still exist in London's Science Museum and in the Conservatoire des Arts et Métiers in Paris.

The same year—1826—Niepce at last succeeded in making the world's first photograph, a view out of the same window where he first pointed his camera in 1816. This time his camera contained a 6½-by-8-inch pewter plate, and he gave it an eight-hour exposure—so long that the sun made an almost complete traverse of the sky, oddly shining on both sides of the structures in the courtyard. Bitumen of Judea was very slow, indeed!

Niepce's first photography of the world outside his window was made with his first professionally built camera obscura, with a meniscus prism made by Charles and Vincent Chevalier of Paris. And it was Niepce's cousin, Colonel Laurent Niepce de Sennecey-le-Grand, who, while on a trip to Paris, went to the Chevaliers' optical shop to buy the camera, and thus caused the first tenuous contact between Niepce and one Louis Jacques Mandé Daguerre, a fast-talking Paris showman who was also avidly pursuing the chimera of photography.

Niepce's cousin had talked a bit too much about the purpose the camera would be put to, and when the Chevaliers raised their Gallic eyebrows he even showed them a heliograph. Daguerre was also a Chevalier customer and the next time he appeared in their shop they told him about Niepce's impressive results.

Daguerre, fearful that others might discover the secret of photography before he did, immediately wrote to Niepce, who at first was very suspicious of his motives, but Daguerre wasn't the kind of man to be put off by rebuffs and kept up his importunities. After a time, Niepce relented and sent Daguerre one of his photoengravings and a printed proof pulled from it. In exchange Daguerre sent him a *dessin fumé*, a sort of drawing in candle smoke which had nothing whatever to do with photography; a fact he didn't tell Niepce. Niepce's engraving hadn't given Daguerre much information either, for Niepce had cagily removed the bitumen coating from it in order to prevent Daguerre's finding out about the chemistry of the heliograph.

Niepce and Daguerre were the results of two entirely different backgrounds. Niepce was a holdover from the *ancien régime*, the son of a king's counselor, an ex-army officer who lived on what was left of the family fortune on the family estate.

Daguerre was, by the stuffy conventions of the 1820s, an upstart. The son of a crier in a magistrate's court, he had learned the technique of overly-realistic painting and used it to extract sous from the pockets of Parisian crowds come to view his dioramas.

Daguerre and a partner, Charles Marie Bouton, invented the diorama in 1821. It was a greatly improved version of those meticulously painted panoramic illusions first produced in England in the late eighteenth century—lamp-lit scenes of cities and fleets and great battles which encircled their viewers sitting in the darkness of cylindrical auditoriums.

Daguerre had been apprenticed to an architect in

3

4

5

1800, when he was only thirteen. At sixteen he started serving another apprenticeship to the celebrated Ignace Degotti, who was chief stage designer of the Paris Opera. In 1807 he became an assistant to Pierre Prévost, the greatest of the panoramists. In 1816 he contracted to become designer-in-chief to the Théâtre Ambigu-Comique, where elaborately realistic and spectacular sets were as important as the melodramatic performances.

Daguerre's sets were more elaborate than any which had been seen before on the Paris stage, but his dioramas made even those seem simple and primitive by comparison. They were not merely beautifully executed three-dimensional painted stage sets. They depended on complex lighting effects from behind through semitransparent sections of the painted scenes, as well as on clever lighting from the front. And they were exhibited in large, especially designed buildings, one in Paris, the other in London. They were sensationally successful.

A London *Times* description of one diorama, "The Valley of Sarnen," suggests their astonishing impact on early nineteenth-century audiences:

"The most striking effect is the change of light. From a calm, soft, delicious, serene day in summer, the horizon gradually changes, becoming more and more overcast, until a darkness, not the effect of night, but evidently of approaching storm—a murky, tempestuous blackness—discolours every object, making us listen almost for the thunder which is to growl in the distance, or fancy we feel the large drops, the *avant-courriers* of the shower. This change of light upon the lake (which occupies a considerable portion of the picture) is very beautifully contrived. The warm reflection of the sunny sky recedes by degrees, and the advancing dark shadow runs across the water—chasing, as it were, the former bright effect before it. At the same time, the small rivulets show with a glassy black effect among the underwood; new pools appear which, in the sunshine, were not visible; and the snow mountains in the distance are seen more distinctly in the gloom. The whole thing is nature itself—and there is another very curious sensation which this landscape scene produces on the mind. The decided effect of the thing is, that you look over an area of twenty miles: the distant objects not included. The whole field is peopled: a house, at which you really expect to see persons look out of the window every moment—a rill, actually moving—trees that seem to wave. You have, as far as the senses can be acted upon, all these things (realities) before you; and yet, in the midst of all this crowd of animation, there is a stillness, which is the stillness of the grave. The idea produced is that of a region—of a world—desolated; of living nature at an end; of the last day past and over."

The fame the diorama brought might have satisfied most men, but Daguerre was insatiable. He wanted a place in history. He knew the diorama was not great art and was perishable, that he would be forgotten when his diorama at last crumbled away. He was, of course, familiar with the camera obscura. Artists had used it for centuries, tracing the images reflected on the ground glass to get correct perspectives. Daguerre must have been fascinated by the camera obscura's images. His continual, almost weekly, visits to the Chevaliers' shop to find ever better cameras and lenses prove as much. Then, in 1824, he burst into the shop, announcing excitedly, "I have found a way of fixing the images of the camera! I have seized the fleeting light and imprisoned it! I have forced the sun to paint pictures for me!"

"Admitting that Daguerre had really found what he announced—and for myself I have no reason to doubt it," Charles Chevalier later wrote, "it is certain that he had cried victory prematurely, or rather, that after having obtained the image he had not been able to fix it." Daguerre, like others, had coated paper with silver chloride and seen only a fleeting, darkening image.

Imagine his terror when he found out that Niepce was also working toward the same end.

"Suddenly, Daguerre became invisible! Shut up in a laboratory which he had arranged in the building of the Diorama . . . he set to work with renewed ardor, studying chemistry, and for about two years lived almost continually in the midst of books, retorts and crucibles. I have caught a glimpse of this mysterious laboratory, but neither I nor anyone else was ever allowed to enter it. . . .

1

2

3

"How many times have I seen him, shut up in his studio for two or three days without leaving it, eating without taking any notice of the food which Madame Daguerre, alarmed, had brought him, not sleeping, obsessed with his idea."

Daguerre did not give up trying to find out how far Niepce had succeeded. He managed to meet him in Paris, where Niepce and his wife had stopped off for passports on their way to London, where Niepce's brother lay ill. Daguerre charmed the Niepces with a tour of the dioramas, but showed them nothing of real photographic importance.

In London Niepce tried without success to enlist the interest of various savants, the Royal Society, and even King George IV, in his heliographs. Low in mind when he returned to France, Niepce was now fair game for Daguerre's importunities. Daguerre pressed him to take views from nature; he was not interested in engravings. But Niepce was still imbued with the idea that views from nature might also be used as printing plates. His engraver suggested that he give up pewter, which was too soft to engrave cleanly, and try copper plates instead. In 1829 Niepce changed his technique to one which would form the basis from which the daguerreotype would directly evolve. He used silver-plated copper plates—the silver would give him the brightness needed for views from nature, the copper the hardness needed for sharp engravings. He still used bitumen of Judea for his sensitized surface. But after exposure and after washing the unexposed bitumen away, he blackened the unexposed parts with iodine vapor. Then he removed the remaining bitumen with alcohol. Now he had a picture in which the light parts were of shining silver, the shadows of darker silver iodide. But the exposure time was not any shorter.

Daguerre, studying one of the new heliographs sent him by Niepce, praised the detail (Niepce was using a sharper lens), but took exception to the harsh tonal values and to the peculiar effects of light and shade caused by the sun shining on both sides of the structures in Niepce's courtyard during the day-long exposure. In his note accompanying the silver-plated heliograph, Niepce told Daguerre that he intended publishing a booklet about his work—"On Heliography." Daguerre, somewhat shaken by this, pressed Niepce to postpone publication. "There should be found some way of getting a large profit out of it before publication, apart from the honor the invention will do you," he wrote.

Niepce was in desperate need of money. He and his brother had sunk most of the family fortune in experiments with an engine-powered boat, the Pyréolophore. Further, he imagined that Daguerre had a marvelous camera, faster and sharper than his, which would enable him to make much shorter exposures. In reality Daguerre had no such apparatus. His camera was the periscopic camera obscura built by Wollaston, the same as that which Niepce had been using, except for having an achromatic lens made by Chevalier, which was hardly better than the lens in Niepce's camera.

Niepce therefore suggested a partnership in which Daguerre would join him in perfecting the heliograph. By the terms of their ten-year partnership, Niepce was to divulge everything about his process. Daguerre, it turned out, had nothing to divulge; he had gotten nowhere with his experiments—no pictures, no remarkable new camera, no process. Niepce had been had.

Niepce, however, held to his commitment to show Daguerre everything about his process. He even made two photographs on glass plates in order to show that his process was not limited to photography on metal. (One of these, of a table set for a meal, lasted until 1909, when the Société Français de Photographie lent it to a Professor Piegnot at the Conservatoire National des Arts et Métiers for tests. While it was in the professor's laboratory he became insanely enraged for some reason and started throwing things around. Sadly, among the items the nutty professor smashed up was Niepce's 1829 still life. A photographic copy of it still exists, however.)

Daguerre had gone by stagecoach to Niepce's home at Chalon-sur-Saône to sign the contract and immediately afterward returned to Paris. He never saw Niepce again, but the partners carried on a considerable and lively correspondence.

Now both Daguerre and Niepce became more in-

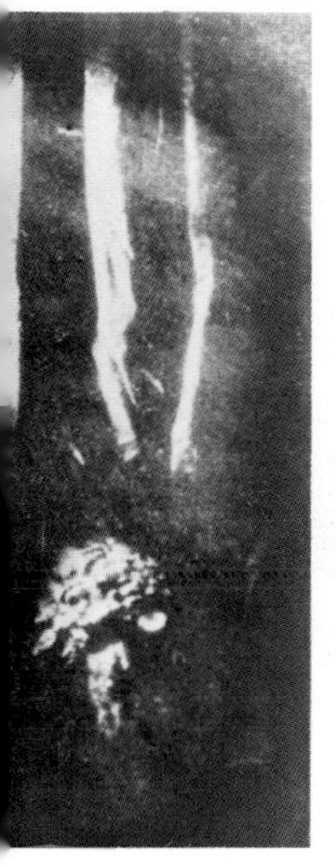

4

1. Painting by Daguerre, about 1829. 2. Daguerreotype of L. J. M. Daguerre, taken by J. J. E. Mayall in 1848. 3. Earliest existing daguerreotype by Daguerre, made in 1837. 4. Stereoscopic daguerreotype of a bust of François Arago, who announced Daguerre's invention of daguerreotypy.

terested in the use of iodine. Niepce, as noted, had been using it to darken parts of his pictures on the silver-plated copper plates treated with bitumen of Judea. Before his partnership with Daguerre he had discovered that iodine on silver (which formed silver iodide) was itself light-sensitive. But he got negative images, which he considered useless, and he knew no way of fixing them. Later Daguerre claimed that he had been first to discover that iodine formed a light-sensitive coating. In any event, after several years of experimenting with iodine and silvered plates, neither Niepce or Daguerre had made much progress.

Suddenly, in 1833, Niepce died of a stroke. Daguerre was on his own. Although Niepce's son Isidore replaced his father as Daguerre's partner, he did very little to carry on the work.

Daguerre continued to drive himself to find some way of getting iodized silver plates to give him pictures. But after exposing them for many hours he still ended up with negative images which he did not know how to fix. In desperation and knowing almost nothing about chemistry, he tried almost every chemical he could lay his hands on. He had some luck turning negative images into positives by the use of carbonic-acid gas, and by exposing his exposed plates to the vapors of petroleum and potassium chlorate. But their tonal gradations were much too crude and contrasty.

Nonetheless, he was on the right track. Niepce had once pointed out that a final strong image was perhaps not necessary, that a very faint image might be strengthened by chemical means after the plate was removed from the camera. The idea of using the vapors from heated chemicals led him to try many substances that would vaporize. The one that finally worked was mercury. The story of how Daguerre discovered that mercury would make visible the latent image as a positive is one of serendipity. In 1835 Daguerre is supposed to have dejectedly thrown an underexposed silvered plate into the cupboard where he kept chemicals, meaning to clean and repolish it in order to reuse it. When he opened the cabinet the next morning he was much surprised to see a picture on the plate.

He exposed another plate for the same time and put it in the cupboard overnight again. Again he had a picture! Day after day he put another plate into the cupboard and each time took out another chemical until only a few drops of mercury remained. Mercury vaporizes at about 68° F. and it was mercury vapor that had performed the miracle.

True or not, it's a nice story.

Daguerre, meanwhile, had not yet hit upon a means of fixing the positive images he could now produce. Still, it did not take him long to start weaseling out of the contract he had made with Niepce. In that contract the firm was called Niepce-Daguerre. Now Daguerre peremptorily announced to Isidore Niepce that a new contract Daguerre had drawn up would style the firm as Daguerre and Isidore Niepce. Isidore—poor, saddled with a wife and four children, and still hoping to profit from the association with Daguerre—gave in and signed the new contract.

In 1835 Daguerre trumpeted that he had found a means of fixing the images. Actually, he hadn't, but two years later he found that a solution of common salt did the job fairly well.

Daguerre again put pressure on poor Isidore. This time he insisted that the iodine-mercury process bear only the name Daguerre, and that it be offered for sale separately from the old Niepce process. Isidore was at first outraged, but again he gave in and signed.

Daguerre now tried to sell the invention to a government, but without success. Later he claimed that England, Russia, Prussia, and the United States, among others, had made him offers which he had refused.

Daguerre had another scheme up his sleeve. He would sell the rights by subscription, and use theatrical tricks to do it. He loaded a carriage with bulky apparatus and ostentatiously pointed his camera at the Louvre, the Seine bridges, and public monuments in central Paris. But the hoopla did him little good. He sold few subscriptions.

Daguerre took another tack. He showed his daguerreotypes to various bigwig scientific types, among them François Arago, a member of the Chamber of Deputies, a director of the Paris Observatory, and permanent secretary of the influential Académie des Sciences. Arago immediately

Right: Three ancient cameras. Wooden sliding-box stereocamera by Ottewill of London is slid sideways on its base to make twin pictures. Brass Voigtländer daguerreotype camera (center) takes round pictures. Small brass French Lerebours daguerreotype camera was used by Fox Talbot.
Below: American Lewis daguerreotype camera of 1851. Opposite: Daguerreotype camera, made in 1839 by Giroux, stands on Fox Talbot's tripod in London's Science Museum.

DAGUERREOTYPE
Paris

recognized the importance of the invention. He felt that it should not be narrowly controlled by a few men, but should freely serve the people of the world. France, he reasoned, should buy the rights to daguerreotypy and give everyone everywhere a free hand to use it. He used all his prestige and power to push the French government toward giving Daguerre and Niepce pensions in return for the rights to the invention.

It was Arago who at last gave the daguerreotype and its progenitor world-wide fame. At the January meeting of the Académie in 1839 he discoursed at length on the wonders of the daguerreotype. Even before this formal and important announcement other notables had seen Daguerre's amazing pictures, and they had talked and written about them enough to set Paris agog.

"From today," the painter Paul Delaroche declared, "painting is dead!"

Among those who had been astounded by the daguerreotype was Samuel F. B. Morse, who was in Paris arranging French patents for his electric telegraph. He had asked Daguerre to let him see his daguerreotypes and offered to show him his telegraph. Daguerre agreed. In a letter to his brother in New York, Morse said, "They [daguerreotypes] are produced on a metallic surface, the principal pieces about 7 inches by 5, and they resemble aquatint engravings, for they are in simple chiaro oscuro, and not in colors. But the exquisite minuteness of the delineation cannot be conceived. No painting or engraving ever approached it. For example: In a view up the street, a distant sign would be perceived, and the eye could just discern that there were lines of letters upon it, but so minute as not to be read with the naked eye. By the assistance of a powerful lens, which magnified fifty times, applied to the delineation, every letter was clearly and distinctly legible, and so also were the minutest breaks and lines in the walls of the buildings, and the pavements of the streets. The effect of the lens upon the picture was in a great degree like that of the telescope in nature.

"Objects moving are not impressed. The Boulevard, so constantly filled with a moving throng of pedestrians and carriages, was perfectly solitary, except an individual who was having his boots brushed. His feet were compelled, of course, to be stationary for some time, one being on the box of the boot black, and the other on the ground. Consequently his boots and legs were well defined, but he is without body or head, because these were in motion.

"The impressions of interior views are Rembrandt perfected. One of Mr. D.'s plates is an impression of a spider. The spider was not bigger than the head of a large pin, but the image, magnified by the solar microscope to the size of the palm of the hand, having been impressed on the plate, and examined through a lens, was further magnified, and showed a minuteness of organization hitherto not seen to exist. You perceive how this discovery is, therefore, about to open a new field of research in the depth of microscopic nature. We are soon to see if the minute has discoverable limits. The naturalist is to have a new kingdom to explore, as much beyond the microscope as the microscope is beyond the naked eye."

Arago, continuing to pressure the French government, succeeded in moving Tannegui Duchâtel, the Minister of the Interior, to draw up a proposal with Daguerre and Niepce. Then a bill could be presented in the Chamber of Deputies and to the king. Under its terms Daguerre was to get a life pension of 6,000 francs (about $12,000 in present-day money) for the right to publish the secrets of the daguerreotype and the diorama. Paying Daguerre for the diorama was a device to give him more than the 4,000 francs Isidore Niepce was offered. The Chamber of Deputies, after being shown a selection of daguerreotypes, passed the bill 237 to 3.

Daguerre was allowed to keep his secret for six more weeks, mostly to give him time for a bit of chicanery—patenting the process in England. This after all the big talk about giving daguerreotypy to all mankind. The hiatus also gave him time to publish a manual and to have apparatus manufactured by Alphonse Giroux, a relative of Mme. Daguerre. Giving the camera-building monopoly to M. Giroux, a stationer and not an instrument maker, caused a certain amount of heartburn for Chevalier. "An optician was

Opposite: Daguerreotype panorama of Paris, attributed to Chevalier. Left: Brady's "Daguerrian Gallery" was more ornate than most photographic emporia. Daguerreotypes were mass-produced for while-you-wait customers until the late 1850s, when the wet-collodion process took over.

needed," he snapped. "A stationer was chosen."

Finally, on a stifling August 19, 1839, the Collège Mazarin of the Institut de France was packed with a sweltering crowd. Every seat was taken by sweating bigwigs in science, literature, and the arts hours before the revelations were to be made. Latecomers filled the corridors, stairways, and courtyard. Outside, on the Quai de Conti, a crowd waited eagerly for scraps of information.

Daguerre and Isidore Niepce sat behind a table on the platform next to Arago. In front of them was ranged a camera, apparatus, and three daguerreotypes. Daguerre failed to speak, pleading a sore throat. Perhaps he feared that questioning from such a distinguished conclave might show up his lack of scientific knowledge. Arago did the talking.

When Arago first described the process the audience was dumbfounded. It had expected weird manipulations with outlandish chemicals. The use of iodine vapor to sensitize the silvered plates and mercury vapor to form the image seemed wonderfully simple. But Arago loved to talk. He went on and on, making the process seem more complicated than it was and annoying Daguerre who began to wish he had spoken after all. Still, cheers rang out when Arago finished. "There was as much excitement as after a victorious battle," wrote one eyewitness. When the lucky members of the audience pushed their way into the mob outside they were begged for information. "Silver iodide," cried one. "Quicksilver," shouted another.

"An hour later," the eyewitness continued, "all the opticians' shops were besieged, but could not rake together enough instruments to satisfy the onrushing army of would-be Daguerreotypists; a few days later you could see in all the squares of Paris three-legged dark-boxes planted in front of churches and palaces. All the physicists, chemists and learned men of the capital were polishing silvered plates."

Few of these first exuberant practitioners of the art were successful. Charles Chevalier and two friends were perhaps the first of those who heard Arago's discourse to get a good picture. But Chevalier was an experienced instrument maker and had quickly built himself efficient apparatus, much better than the homemade, spectacle-lensed cameras and cigar-box vaporizers built by eager amateurs. Few people were able to get hold of the "official" cameras built by Giroux. For despite their high price of 400 francs ($800 in today's money), they sold out in a few hours.

There was considerable carping in the press that daguerreotypy was not as simple as Arago had made out, that it was an art impossible for most people to practice. Outraged by the press attacks, Daguerre told one of the editors, Jules Janin of the magazine *L'Artiste*: "You are wrong. True, my process requires a certain amount of care, but anybody can do it . . . come right away to the fourth floor of my house, and before your eyes I'll make a picture as exact, as true, as brilliant, and as beautiful as Raphael himself could ever have made."

Janin, as he later reported in his magazine, accepted the offer. "We arrived at Daguerre's and, not without emotion, went into the little room where he works his marvels. . . . Daguerre's studio is very simple: on the walls are pretty engravings, mediocre plaster casts, retorts and beakers. . . . M. Daguerre had already put the box for the iodine on the table. The mystery was about to begin.

"Daguerre took a lightly silver-plated copper plate. He poured some acid on the silver. He wiped it dry, then rubbed on it a little powdered pumice stone that he moistened with acid. That done—and it is very simple—he attached a thin border of the same metal around the plate with some screws which were in readiness. The plate was now put on the iodine box. The iodine is on the bottom of the box and throws its vapor through gauze upon the mirrored surface of the plate. The window shades of the room were drawn, it is true, but the darkness was not so great that we could not see each other quite well. From time to time Daguerre took the plate from its box; and not finding it sufficiently covered with iodine, he put it back in its place until finally the iodine was spread equally over the surface, which took on the color of gold. This operation takes scarcely a quarter of an hour. That done, you place your colored plate in a kind of wooden portfolio. The camera awaits you in the next room. You choose the view you want to reproduce and then you put in

1

2

3

4

'Latticed Window
(with the Camera Obscura)
August 1835

When first made, the squares
of glass about 200 in number
could be counted, with help
of a lens.

5

6
7

1. William Henry Fox Talbot, 1864. 2. Fox Talbot's camera obscura. 3. Solar microscope, used by Talbot to obtain photomicrographs. 4. One of Talbot's photomicrographs of insect wings, about 1841. 5. Oldest negative in existence, made by Fox Talbot in August, 1835. 6. Three cameras used by Fox Talbot, 1835-39. 7. Talbot's cameras of 1840-42.

the camera your iodized plate, without opening the case which protects it. Once in the camera, the case is opened by a little spring, and soon the prodigy begins. Light coming from everywhere throws on the plate all its power and life. The exterior world is reflected in the miraculous mirror. At this moment the sun was lightly veiled. 'We'll need six minutes,' Daguerre said, pulling out his watch. And indeed, at the end of six minutes he closed the box in which the plate was contained, and on the plate, all the beautiful landscape invisible to the eye. Now all he had to do was to say to this hidden world: *'Show yourself.'* Another box was prepared. It contained mercury. By means of a lamp, this mercury is heated until it reaches fifty degrees [Centigrade], then little by little, through a glass put in the box expressly for that purpose, you see the all-powerful vapor mark each part of the plate in the appropriate tone. The landscape appears as if it had been drawn by the invisible pencil of Mab, the queen of the fairies. When the work is done, you take out the plate, and put it in hyposulphite,* after which you throw warm water over the plate. The operation is finished, the drawing is whole, complete, unalterable. All that in an hour, more or less. . . .

"Thus this operation, which seemed almost impossible as recounted by M. Arago, is very easy and simple, as done by Daguerre."

People came by the hundreds to learn the process from Daguerre—"the master"—who gave demonstrations in a huge room in the palace of the Quai d'Orsay. They were further aided by a booklet he published on daguerreotypy, which became a world-wide best seller in a matter of weeks. Tailors, printers, butchers, dentists, men of every station and occupation in almost every country became instant, maniacal daguerreotypists. No invention before or since has ever been such a wild sensation.

Nowhere was it as much of a success as it was in the United States. The first American daguerreotype picture was made on September 16, 1839, by D. W. Seager, an Englishman living in New York. Quick work indeed, since Daguerre's directions were only published on August 20. Seager, who was abroad at the time, seems to have been tossed a copy of the manual just as the fast steamer on which he was sailing to America was leaving its dock. Seager not only displayed his daguerrian view of St. Paul's Church in Chilton's drugstore in New York, but gave lectures and demonstrations, and compiled the world's first exposure table, which was included in the first American booklet on photography published by druggist Chilton in March, 1840.

Soon there were many American experimenters, including Samuel Morse, Dr. John Draper, and Alexander Wolcott, who invented a camera which worked on the principle of the reflecting telescope, using a concave mirror instead of a lens. With this faster camera Wolcott opened the world's first portrait studio. Previously exposures had taken five minutes or more, too long a time for any sitter to hold a pose.

It was not only faster-working cameras and lenses that made portraiture practical. In 1841 Antoine Claudet, a Frenchman living in London, phenomenally increased the speed of the daguerreotype plate by fuming it with chlorine vapor, as well as iodine, and making it possible to take portraits outdoors, in the shade, in under ten seconds, and as quickly as one second in sunlight. American daguerreotypists using bromine vapor in addition to chlorine, and various other combinations called "quick stuff," soon made the process even faster.

By 1841 almost every American town had its resident daguerreotypist. Those that didn't were visited by itinerant operators. By the end of the decade there were some ten thousand American daguerreotypers at work. In New York, on lower Broadway, a hundred "daguerrian galleries" mass-produced portraits. Some three million were turned out in the United States each year during the 1850s and competition forced prices down to as little as twenty-five cents for a small one in a fancy imitation-leather case.

Yet by the late 1850s the daguerreotype was stone-cold dead in Europe, and before the Civil War had ended it was almost gone in America, too. That other inventor of photography, William Henry Fox Talbot, had planted the seed which grew into the process that killed the daguerreotype. Fox Talbot was the archetypical moneyed Vic-

*Until March, 1839, when Sir John Herschel told the Royal Society about "hypo," Daguerre had used table salt.

Calotypes by Fox Talbot.
1. & 2. Talbot's printing establishment at Reading, about 1844.
3. Trafalgar Square, with the Nelson column under construction.
4. Copper ore vessel.
5. Miss Horatia Fielding, Talbot's half sister.
6. "Two Figures."
7. "Woodcutters."

1

3

2

5

4

6

7

Calotypes by Adamson and Hill.
1. The Reverend James Fairbairn and Newhaven fishwives, 1843.
2. John Henning and Alexander Handyside Ritchie, about 1843.
3. Mrs. Anna Brownell Jameson, about 1846. 4. Newhaven fisherwomen, about 1845.

1

2

torian gentleman, the grandson of an earl, the son of an officer of dragoons and his titled wife. Unusually for a member of his social class Talbot was no layabout. He did things.

After Harrow and Trinity College, Cambridge, and papers on the classics and maths for which he was elected a Fellow of the Royal Society, he stood for Parliament as a Liberal from Chippenham and was elected in 1833, when he was thirty-three.

But Talbot was bored with politics and spent more time traveling on the Continent than he did in the House of Commons. During these excursions abroad he tried tracing the contours of landscapes he admired on sheets of paper on the screen of a camera obscura. But even tracing requires some artistic ability and poor Talbot had neither the eye nor the hand for it. After a trip he made to the Italian lakes, he returned to England determined to find another way. "How charming it would be," he wrote, "if it were possible to cause these natural images to imprint themselves durably, and remain fixed upon the paper." Talbot had no idea that a devious little French scene painter named Daguerre was feverishly working toward a similar end.

He started as Wedgwood had, by bathing writing paper in silver nitrate and then exposing it to light. This darkened, but much too slowly. He then tried silver chloride formed by first coating his paper with a solution of ordinary table salt (sodium chloride), letting the paper dry, and then bathing it in silver nitrate. This darkened slowly, too, and Talbot experimented with various strengths of both solutions. He found a weak sodium-chloride solution best; the paper blackened rapidly and evenly in daylight. He discovered, too, that a strong solution of salt water acted as a fairly efficacious fixer.

In the beginning Talbot's pictures were only contact prints of things like lace and leaves, but after he had succeeded in fixing these images he tried to take pictures in the camera. At first his sensitized paper was much too slow and he tried bathing it alternately in salt water and silver nitrate, and exposing it wet in a big homemade box camera. After an exposure of several hours in summer sunlight it was

3

4

still much underexposed. The pictures he got were mere silhouettes without detail.

Talbot decided that the long-focus "object lens" of his big camera was too slow. He had some very short-focus microscope lenses of fairly large diameter. To accommodate their short focal length he had local cabinetmaker build him some tiny cameras two and a half inches square—"little mousetraps" as his wife called them. With these he got "very perfect but extremely small pictures . . . as might be supposed to be the work of some Lilliputian artist." The London Science Museum still has one of these inch-square paper negatives. It is of a many-paned lattice window taken from inside Talbot's country seat, Lacock Abbey. On the card upon which Talbot carefully mounted it is written: "Latticed window (with the Camera Obscura) August 1835—when first made the squares of glass about 200 in number could be counted with help of a lens." This stamp-sized bit of paper is the oldest negative in the world.

It was sometime between 1835 and 1839 that Talbot got the idea that he could make positive photographs by letting the sun shine through a paper negative onto a sensitized sheet of paper—an obvious concept which seems not to have occurred to other experimenters. But it still hadn't occurred to Talbot that after-treatment—development—of the latent image would immeasurably speed up the process. He still hoped that it would be possible to open a camera and take out a fully finished picture.

Fox Talbot, secure in his social position and interested in other scientific studies besides "photogenic drawing," as he called it, was not, like Daguerre, trying desperately to claw his way to success and fame. He had made no attempt to publish his results. Yet, when Arago announced Daguerre's discovery at the January meeting of the Académie, Talbot seems to have turned a gentlemanly purple. He immediately wrote Arago and Daguerre's friend, Biot, claiming priority for his invention. And he managed to get himself some publicity. At one of the regular Friday night meetings of the Royal Institution, the great Michael Faraday announced that Talbot had made a new discovery. A small display of photogenic drawings was hung on the walls of the Institution's library. Talbot himself read a paper to the Royal Society, entitled "Some Account of the Art of Photogenic Drawing; or the Process by which Nature's Objects may be made to Delineate Themselves without the Aid of the Artist's Pencil."

These meetings were important. But more important was the work of a man who couldn't attend because of a "rheumatic affection." This was the famous scientist, Sir John Herschel, who had heard of Talbot's and Daguerre's work, and without knowing the details of their processes proceeded to make his own photogenic drawings on paper. He fixed them with "hypo," then known as hyposulphite of soda and now called sodium thiosulphate. To this day, a bath of "hypo" is our method of removing unexposed silver salts and thus fixing negatives and prints. It was one of the most important of all photographic discoveries.

Daguerre, who, like Talbot, had been using common salt to fix his daguerreotypes, immediately switched to hypo. But Talbot persisted for several years in the use of salt, which resulted in rapid fading of his images.

The great excitement over the public disclosure of Daguerre's process seems to have moved Talbot to get cracking on improving his photogenic drawings. In 1840 he, like Daguerre, discovered the latent image. He found that he didn't have to wait for the sensitive sheet of paper in his camera to show an image, but that he could remove it while still blank and then *develop* it. The image would, as Talbot wrote, "appear spontaneously."

This tremendous discovery shortened exposures from over an hour to a few minutes. Talbot named it the calotype process ("beautiful" in Greek). He first sensitized a sheet of paper in gallo-nitrate of silver (gallic acid, silver nitrate, and acetic acid). After exposure he developed it in the same solution. Then he fixed the paper, washed it, and used it as a negative to make prints which were developed in the same manner.

Talbot had heard about gallic acid in April, 1839. A Rev. J. B. Reade discovered that it increased the speed of photogenic drawing paper and told Andrew Ross, an optician, about it. Ross, who made lenses for both Reade and

1

2

3

Talbot, told Talbot. Reade seems never to have tried gallic acid as a developer.

In the 1840s and 1850s the calotype was used mostly for views—for landscape and architectural photography. The daguerreotype still held the lead in portraiture. In spite of the popularity of the silvered copper plates, two Scotsmen, David Octavius Hill and Robert Adamson, made some of the most magnificent photographic portraits of all time in calotype. Hill was a painter; Adamson a twenty-two-year-old photographic technician. Hill was engaged in painting a huge canvas portraying the four hundred and fifty delegates to the conclave at which the Free Church of Scotland was founded. He decided to work from photographs of this crowd of ministers and approached Adamson, who was working the calotype process in Edinburgh.

Not only the churchmen came before their cameras. They photographed fishwives, nobility, academics, fishing boats, cottages. In all they made some fifteen hundred negatives before Adamson died at the age of twenty-seven in 1847.

Perhaps the very limitations of the calotype process helped make the Hill-Adamson portraits so great. The long exposures outdoors in sunlight forced their subjects into attitudes of quiet, dignified repose. And the lack of detail helped produce masterly compositions in broad masses of tone. But Hill and Adamson, especially Adamson, were geniuses nevertheless.

There were, of course, some drawbacks in the calotype process. (Talbot's friends insisted it be called Talbotype.) The grain of the paper negative was somewhat obtrusive in the prints. Further, in the early years the prints tended to fade. Nor did the calotype have the biting sharpness and brilliancy of the daguerreotype. Yet the calotype had one tremendously important advantage: Any number of prints could be made from one negative. In fact, prints are still being made from Talbot's negatives stored in the London Science Museum. The daguerreotype was a one-shot process. To make a duplicate it was necessary to rephotograph a daguerreotype. The calotype was also far less fragile. It could be mailed, pasted into an album, retouched.

Despite the fact that it did not have the popular success of the daguerreotype, the calotype evolved into the negative-positive photography of our time. The daguerreotype came to a dead end.

By the mid-1840s, then, photography had not only been invented, it had become practical and was in use in every part of the world. Thenceforth every advance would merely amplify the ideas of Niepce, Daguerre, and Talbot.

In 1851, the year Daguerre died, the wet-collodion process ended the era of the daguerreotype and calotype. Negatives on glass combined the advantages of both earlier processes, the daguerreotype's sharp clarity and the calotype's ability to make an infinite number of prints from a single exposure. And the process needed less exposure time than either of the others. It was the invention of Frederick Scott Archer, an Englishman. For thirty years, until the advent of the gelatine dry plate in the late 1870s, it was virtually the only method used for making photographic negatives.

But working the collodion wet-plate process was an enormous chore, especially for photographers away from their home studios and darkrooms. Iodized collodion had to be poured onto a glass plate (a 10-by-12-inch plate was considered medium size; many were larger). Immediately thereafter it was sensitized by dipping it into a solution of silver nitrate. The exposure had to be made while the plate was still wet—for collodion lost its sensitivity as it dried and crystallized. Development, too, had to be done before any drying took place. Pity the poor traveling photographer who had to carry his darkroom—his bottles and trays and baths—with him into the field either in the form of a darktent on his back, or on a pushcart, or aboard a wagon, as did Englishman Roger Fenton, the photographer of the Crimean War, and Mathew Brady in the American Civil War.

To obviate the traumas of the wet-plate process, various means of preserving the sensitive collodion coating were tried. And later dry collodion plates of exceeding slowness had only moderate success in the late 1860s.

The gelatine silver-bromide emulsion, first coated on dry plates and later on film, caused a revolution in photography. Dr. Richard L. Maddox, of England, is usually

4

1. The Reverend J. B. Reade.
2. Frederick Scott Archer, inventor of the wet-collodion process.
3. Wet-collodion photographer's dark-tent mounted on a pushcart.
4. Wet-plate developing camera designed by Archer in 1853. A glass plate was sensitized and developed inside the camera through cloth sleeves in order to dispense with the usual dark-tent. The tent was more practical.

given credit for the invention of the dry plate. But other men, notably John Burgess of London, made the new process practical. By 1876 it was possible to buy a box of dry plates from the Liverpool Dry Plate Company, load them into plateholders, and go forth with a camera to take pictures unencumbered by the wearisome paraphernalia of the wet-collodion process. By 1882 wet plates were as dead as the daguerreotype and the age of the amateur had begun.

Now almost anyone could take photographs, and manufacturers hastened to build small cameras which could be held in the hand. Dry plates were so fast that tripods could be left at home. In 1888 the Kodak appeared, with the slogan, "You press the button, we do the rest." We have been pressing that button ever since.

For half a century after Nicéphore Niepce coated a sheet of pewter with bitumen of Judea and exposed it all day, almost every advance in photography was usually the work of a lonely experimenter who worked out some new device or process while quietly tearing his hair in his laboratory.

In more recent times, however, as in most other fields, new ways in photography were devised in the research laboratories of giant corporations by groups of men, each of whom might be a specialist in a single abstruse branch of the art. Over the years these gentlemen made great strides. Film became faster, film grain became smaller, cameras became more convenient, lenses ever faster and sharper. There were, however, two notable exceptions to this mode of corporate improvement in photography. One was Oscar Barnack's invention of the 35mm Leica camera in 1914, which, when first marketed eleven years later, revolutionized photography once again. The second was the invention of Kodachrome in the early 1930s by two noncorporate experimenters, the musicians Leopold Godowsky, Jr. and Leopold Mannes. In both these instances, however, big companies took over the inventions to improve and market them.

Dr. Edwin H. Land's SX-70 automatic color photography system was arrived at in a new and different manner. In a way it was not an invention but a daring conception—a conception seemingly impossible to conventional photographers and photographic chemists and engineers, who had concluded that the instant photographic print Niepce, Daguerre, and Talbot had hoped for was not possible. Those pioneers, in their ignorance of the formidable problems in their paths, hoped that it might be possible to point their cameras at landscapes or persons and record the beautiful images they saw in their ground glasses immediately and in full color. They settled for very much less, and for more than a century all those who followed them settled for less.

But not Dr. Land. He once said, "If you are able to state a problem—any problem—then the problem can be solved." He gave himself the problem of instant photography in 1943, when his three-year-old daughter asked him why she could not see the picture he had just taken of her. Land and his family were on vacation in Santa Fe at the time. "As I walked around that charming town," he recalled later, "I undertook the task of solving the puzzle she had set for me. Within the hour the camera, the film and the physical chemistry became so clear that with a great sense of excitement I hurried to the place where a friend was staying, to describe to him in detail a dry camera which would give a picture immediately after exposure. In my mind it was so real that I spent several hours on this description. Four years later we demonstrated the working system to the Optical Society of America. All that we at Polaroid had learned about making polarizers and plastics, and the properties of viscous liquids, and the preparation of microscopic crystals smaller than the wavelengths of light was preparation for that day in which I suddenly knew how to make a one-step photographic process. I learned enough about what would work in enough different fields to be able to design the camera and film in the space of that walk."

The first result of that walk around Santa Fe was the 1948 Model 95 Polaroid camera, which made sepia prints. In the years since, improvements, new cameras, new films (including color), professional films, and dozens—hundreds—of new products and devices relating to what Dr. Land called "one-step photography" have flooded out of the Polaroid laboratories. Since 1948 some twenty-seven million Polaroid cameras have been sold.

1. Roger Fenton's photographic van in the Crimean War. 2. "Harbor at Balaclava," Crimean War wet-collodion photograph by Fenton. 3. "Mrs. Herbert Duckworth," 1868 wet-collodion photograph by Julia Margaret Cameron. 4. & 5. Faster dry plates, which displaced the wet-collodion process, required exposure meters and shutters to give approximately correct exposures. These date from the 1890s.

1

2

3

4

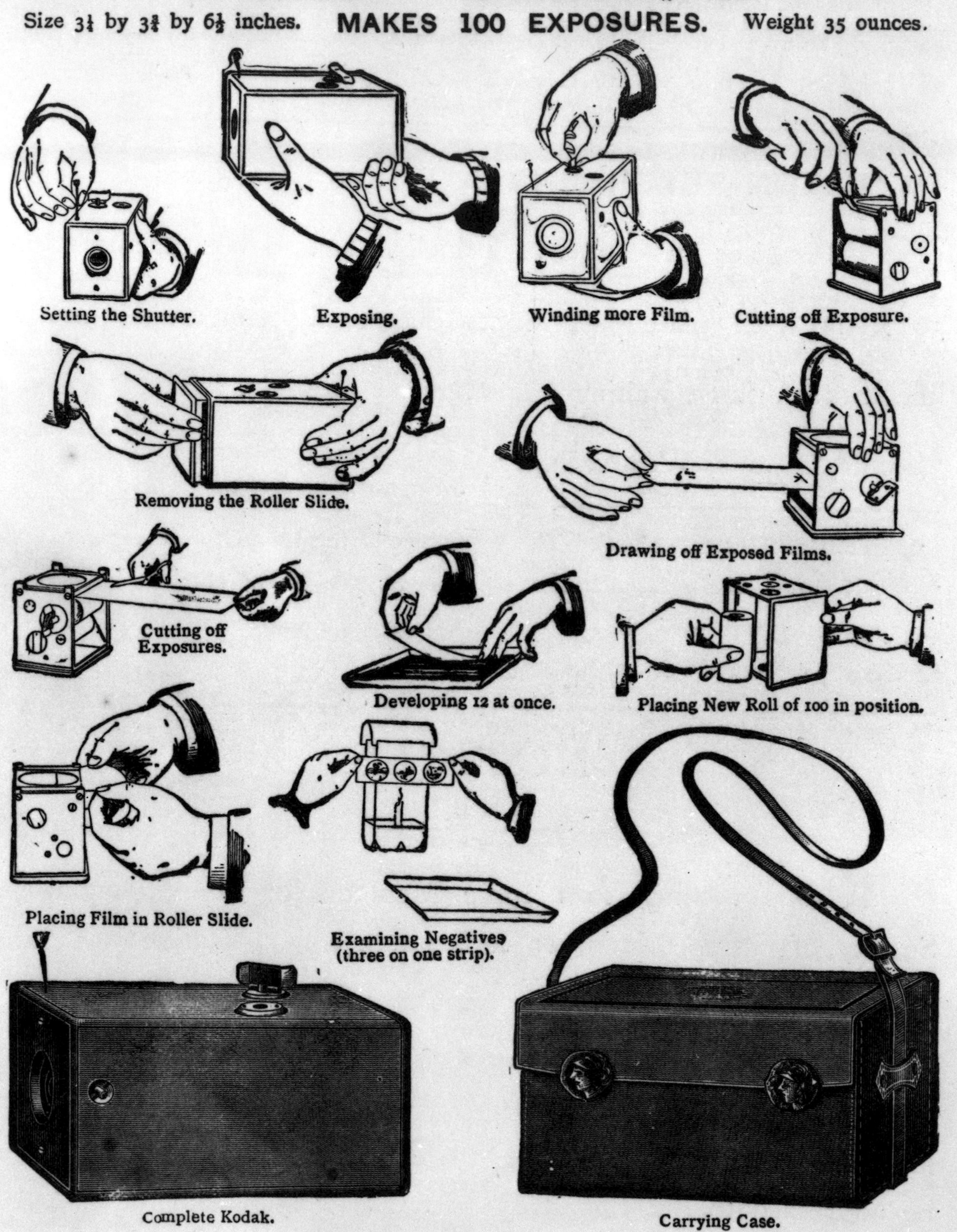
THE KODAK
Is the smallest, lightest, and simplest of all Detective Cameras—for the ten operations necessary with most Cameras of this class to make one exposure, we have only 3 simple movements.
NO FOCUSSING. NO FINDER REQUIRED.
Size 3¼ by 3¾ by 6½ inches. MAKES 100 EXPOSURES. Weight 35 ounces.
Setting the Shutter.
Exposing.
Winding more Film.
Cutting off Exposure.
Removing the Roller Slide.
Drawing off Exposed Films.
Cutting off Exposures.
Developing 12 at once.
Placing New Roll of 100 in position.
Placing Film in Roller Slide.
Examining Negatives (three on one strip).
Complete Kodak.
Carrying Case.
FULL INFORMATION FURNISHED BY
THE EASTMAN DRY PLATE & FILM Co., 115, Oxford St., London, W.

1

1. Early advertisement for the Kodak. 2. George Eastman using a No. 2 Kodak aboard the S.S. *Gallia*, 1890. An associate with a similar camera took Eastman's picture. 3. First Leica model to be sold to the public in 1925. This one bears serial number 518. 4. First Polaroid Land camera, 1948. 5. Recent Polaroid SX-70 camera.

But Dr. Land had not reached his goal. What he aimed for—"*absolute* one-step photography"—was finally achieved in the almost magical SX-70. Users of earlier Polaroid cameras had the not-too-onerous task of taking the picture, waiting some seconds for it to develop, peeling the negative and positive apart, and then finding some place to throw away the disgustingly gooey negative.

A million or so owners of the SX-70 now know that all they have to do is "press the button, the *camera* does the rest." The pictures zip out, clean and dry, about as quickly as the button can be pressed. The button presser can enjoy watching the images develop in full color in bright sunlight. Particularly pleasing to ecology-minded Dr. Land, there are no sticky negatives left over to defile the landscape.

But how does Land do it? How does he invent these marvels? The answer is that he doesn't. What he does is more remarkable. He conceives and defines his revolutionary ideas whole. They are complete in his brain. He then has hundreds of scientific types—chemists, physicists, electronic engineers, optical engineers, and kinds of engineers only other engineers have ever heard of—make the ideas work. After ten years and about a quarter of a billion dollars, Dr. Land had his SX-70.

Dr. Land doesn't see his investigations into one-step photography as merely scientific or technical. As he said in an address before the Royal Photographic Society: "The purpose . . . is essentially aesthetic . . . to make available a new medium of expression to the numerous individuals who are not given to drawing, sculpture, or painting."

And wasn't that old Fox Talbot's problem when he tried tracing those Italian landscapes on his camera obscura in 1833?

Right: Seventeenth-century magic lantern by Kircher. Opposite: Nineteenth-century toy magic lantern was a precursor of home movies. Pictures were painted in color on glass slides; the projector was lit by kerosene.

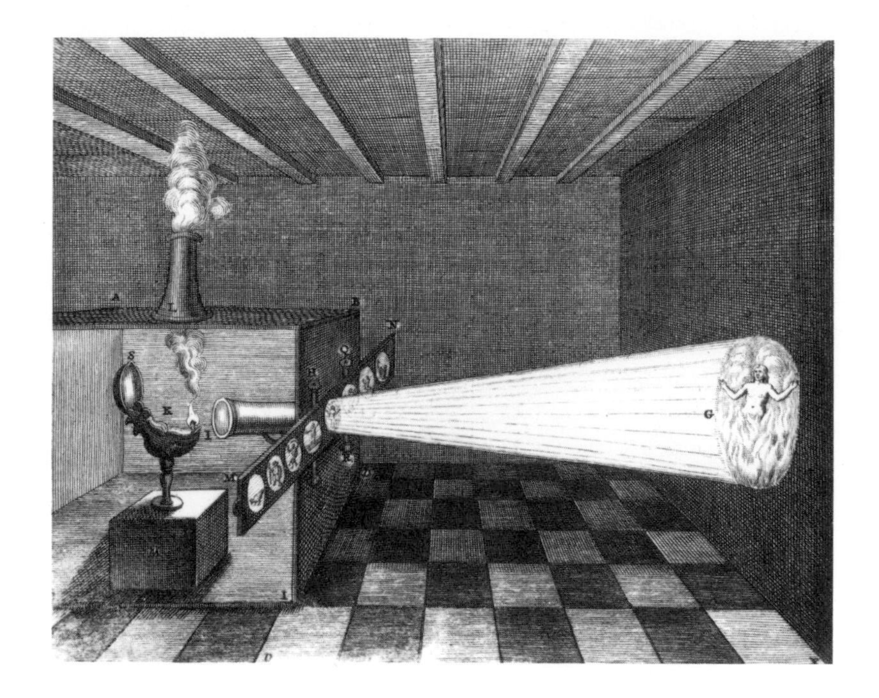

The Animated Picture

Although there are other claimants to the honor, it is generally agreed that the Lumière brothers, dry-plate manufacturers of Lyons, France, were the first to show moving pictures on a screen to a paying audience. This they did at the Grand Café in Paris in December, 1895. The movies were short; film was not yet made in long lengths. For fifty seconds or so the audience could watch Lumière workers coming through the factory gates, a fishing boat docking, a train coming into a station, and so forth. The subjects were commonplace, but audiences were enthralled by the fact that photographs could be made to move.

Like most inventions, moving pictures depended on previous discoveries: (1) the magic lantern; (2) a bright-enough light for the lantern—the electric arc light; (3) celluloid film upon which to coat the negative and positive emulsions; (4) the gelatine-bromide process of photography.

Before photography was invented there had been optical toys in which hand-drawn figures and animals gave the illusion of motion due to the phenomenon of the persistence of vision. They all depended on revolving discs or cylinders upon which were printed the sequential drawings viewed either through revolving slits or mirrors. And they had wonderful nineteenth-century pseudoscientific names: phenakistascope, zoetrope, praxinoscope. But toys or not, they gave early movie pioneers some insight into what was required in shutter design and helped them decide on the number of pictures per second that would result in smooth action.

Eadweard J. Muybridge, who took the first series of action photographs, used a development of one of these optical toys to project his photographs. He called his device a "zoopraxiscope." In 1872 Muybridge had been commissioned by Governor Leland Stanford of California to photograph his famous trotter Occident to determine whether it ever had all four feet off the ground at the same time. But his wet-collodion plates were too slow to stop the action. Muybridge tried again in 1877, this time with a specially designed 1/1000-of-a-second shutter and a hypersensitized collodion emulsion on his plates. This time he got a silhouette of the horse against a white background and proved that it did at a certain point in its stride lift all four hoofs off the ground. The picture created a furor and infuriated some sportswriters. They called it a downright fake.

With Stanford's backing to the tune of $40,000—a lot of money in 1878—Muybridge then went on to analyze photographically the motions of a trotting horse in rapid action. He first set up a battery of twelve cameras whose shutters were electrically tripped by wires across the track. In 1879 he set up twenty-four cameras. Now he had twenty-four frames of a moving picture, and if the glass plates could have been printed on a strip of modern 35mm film he would have had a one-and-a-half-second movie of a trotting horse. But such film was far in the future.

Anyhow at first Muybridge did not think of projecting his pictures. But after they were published in magazines in the United States and Europe, several writers pointed out that if the pictures were mounted on the paper disc of a zoetrope the horse could be seen in motion. In fact, *The Field,* a London sporting magazine, did exactly that and attracted crowds to its show window with a motorized zoetrope.

Muybridge went them one better with his zoopraxiscope with which he could project his pictures in animated form. He mounted the photographs around the periphery of a glass disc. Another opaque disc, containing the same number of slots as there were pictures of the horse, was mounted in front of the glass disc. When both discs were spun in opposite directions in front of a magic lantern, the audience saw a running horse on the screen.

In 1881 and 1882 Muybridge created a sensation demonstrating his zoopraxiscope and its images of horses, deer, birds, and athletes before learned societies in Europe. In London he lectured to the Royal Society, the Royal Academy, and other similar bodies of boffins. He elevated what had been an amusing toy for children to a serious sci-

1. Replica of Muybridge's zoopraxiscope. 2. Nineteenth-century lecturer using a limelit triple lantern for dissolving views.
3. Phenakistascope discs.
4. Praxinoscope theater of the 1870s.
5. Praxinoscope and candle.
6. Eadweard J. Muybridge.
7. Muybridge's arrangement of twenty-four cameras and trip-wires for photographing trotting horses. The cameras were in the white building. 8. One of Muybridge's studies of animal locomotion.

1

2

3

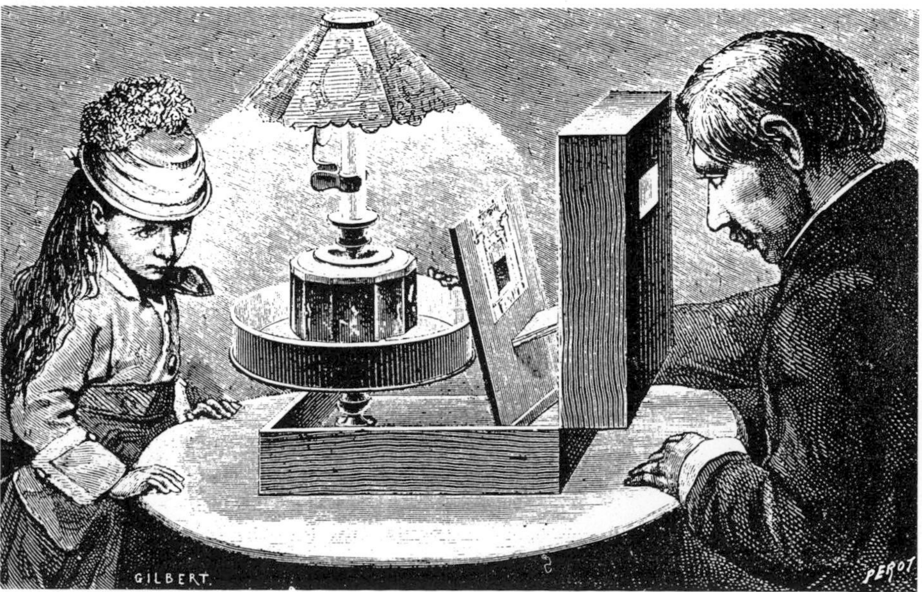

4

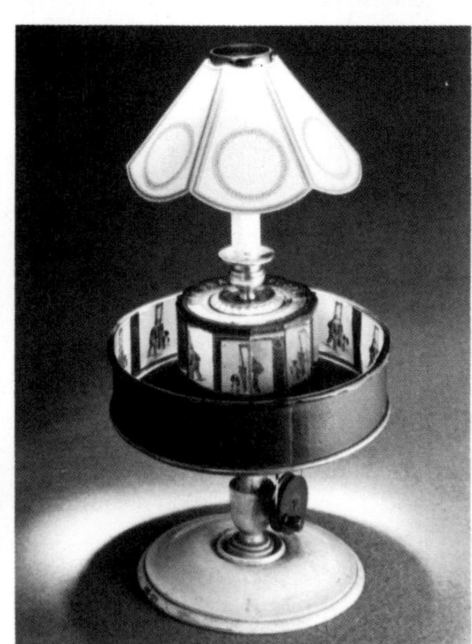
5

6

7

8

1. Marey's photographic gun of 1882. 2. By 1890 Marey was using this moving-picture camera. 3. Multiple images of a life-size wax model of a seagull inside a large zoetrope. 4. Replica of Rudge's lantern. 5. William Friese-Greene. 6. Stereoscopic movie camera built by Friese-Greene and an associate named Varley. 7. Stereo film taken in Hyde Park, London, by Friese-Greene.

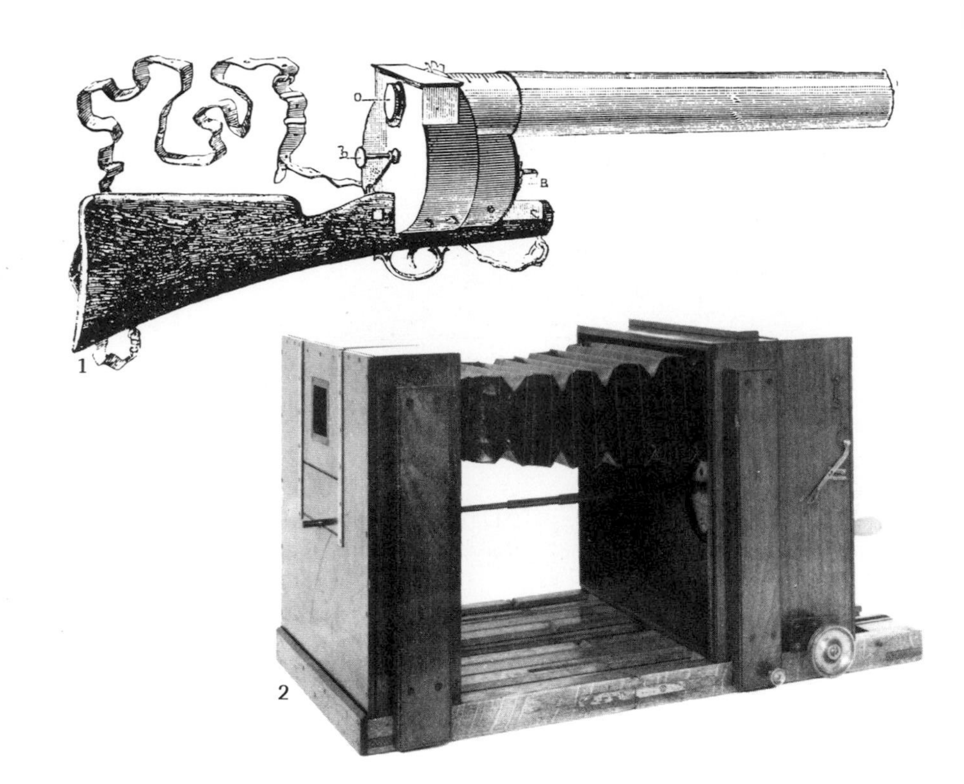

entific instrument for the study of animal locomotion. More than that he had inspired a host of inventors to work on moving pictures.

Professor E. J. Marey, a French physiologist with a great interest in the study of animal movement and bird flight, had met Muybridge during his tour. Muybridge's kind of photographic analysis would, he felt, suit him perfectly. But obviously Muybridge's photographic system—the wires, the twenty-four cameras, and so forth—wouldn't work at all with flying birds. A single gun-like camera he could aim at a bird while shooting a burst of pictures was what he needed.

In 1874 a French astronomer, P. J. C. Janssen, had taken photographs of the transit of Venus with such a camera, albeit a much slower one than Marey needed. Janssen took forty-eight pictures on a circular plate and each exposure took some seventy seconds. The plate was automatically turned by clockwork. Marey's gun camera was based on Janssen's, but it banged off twelve exposures of postage-stamp size in one second at 1/720 of a second per exposure. In February, 1882, Marey first tried it out successfully on a seagull silhouetted against the sky.

In 1887 Marey built what he called a "Chronophotographe." This was the world's first true motion-picture camera. He used rolls of sensitized paper (until rollable, transparent celluloid film became available in 1890). The pictures it took were big by modern standards—nine centimeters (three and three-fifths inches) square. The film was spooled within two magazines and was stopped at the moment of exposure by an electromagnet. Marey's film was unperforated and the spacing between exposures was uneven, making it almost impossible for the film to be projected after it was printed. As many as sixty exposures a second were possible at a shutter speed of 1/1000 of a second. But in Marey's day the longest roll of film available was only four meters long.

Marey's chief interest was the study of animal movement and he had little interest in public exhibitions, but by 1892 he succeeded in equalizing the spacing between frames and then, using an arc light, projected very short studies of action on a screen. Marey not only built the first cinema camera, but he also was the first to use two devices still in use today: the loop in the film just before it gets to the gate, and the leader, that opaque length of paper or film which makes it possible to load a roll of movie film into a camera in daylight.

At about the same time that Marey was experimenting with chronophotography, an English inventor and photographer, William Friese-Greene, was similarly involved in preparing a series of lantern slides for an unusual projection lantern devised by a mechanic named Rudge. Rudge's lantern showed seven glass slides in rapid succession when the operator cranked a handle. The crank also operated a ground-glass shutter which partly blocked off the light while each slide moved into position behind the lens, thus giving a dissolving effect. This lantern, which Rudge called a "Bio-Phantascope," was patented in 1884. This rather primitive device seems to have inspired Friese-Greene, after Rudge's death, to construct what he called a "machine-camera." With a friend, Mortimer Evans, an engineer, he succeeded by 1889 in building a camera which took ten three-and-a-quarter by four-and-a-quarter-inch pictures per second on rolls of perforated, sensitized paper which he oiled in order to make prints. He used celluloid film when it became available.

There is no evidence that Friese-Greene ever publicly projected moving photographs taken with his "machine-camera." He was supposed to demonstrate his apparatus at the Chester meeting of the Photographic Convention of the United Kingdom in June, 1890, but all he was able to show was his camera and a long strip of developed film. His projector seems to have been damaged on the way to the convention. Still, he was at least five years ahead of the procession of moving-picture inventors in using perforated film moved intermittently, frame by frame, with a shutter hiding each movement of the film. Friese-Greene and Evans patented their mechanism in 1889, and years later, in 1912, when their patent was challenged by Edison's Motion Picture Trust, the United States Supreme Court upheld the Friese-Greene patent as the master patent of the world in cinematography.

Many other experimenters in the late 1880s and

3

4

5

6

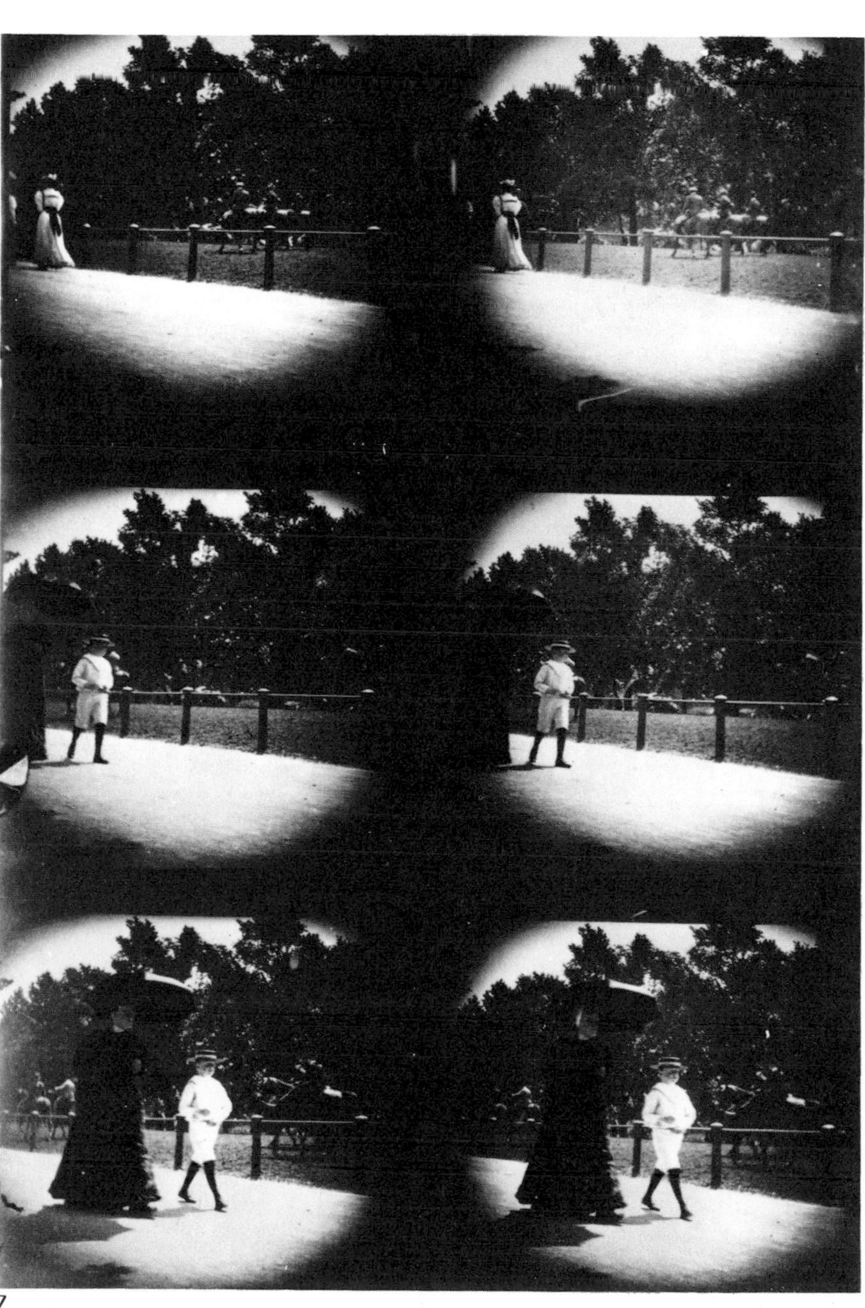

7

O
E
S
R
V
P
POYET

POYET

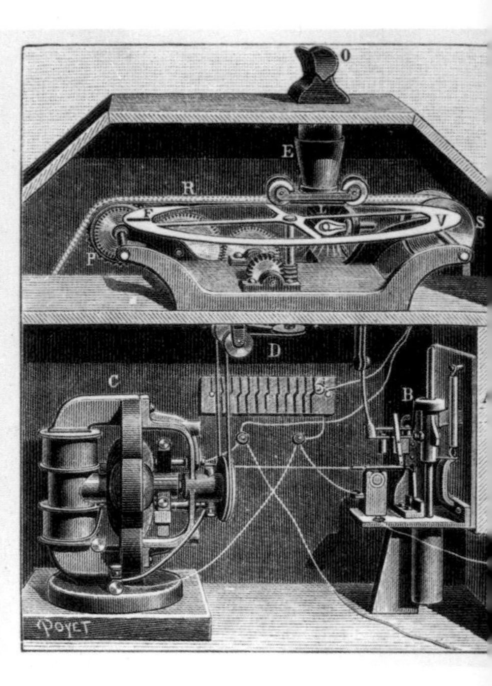
O
E
R
P
V
S
D
C
B
POYET

1

POYET

2

CINÉMATOGRAPHE

3

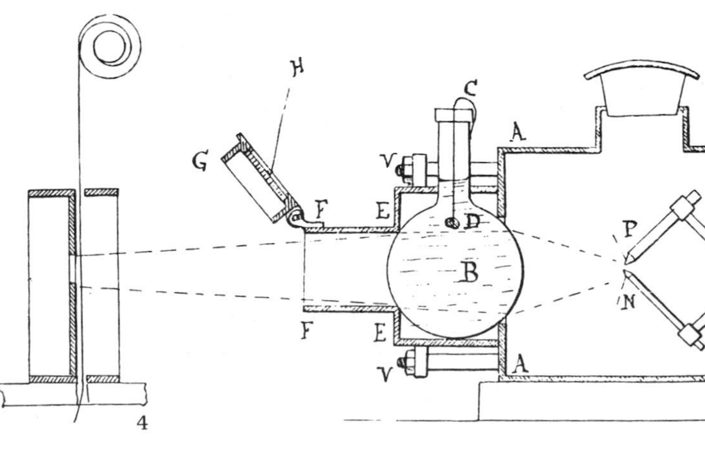
H
C
G
V
A
F
E
D
P
B
N
F
E
V
A

4

1. Two Edison kinetoscope models. 2. Lumière projectionist, 1896. The film falls into a box instead of being rewound. 3. The Lumière camera also served as a projector. 4. Diagram of the Lumière projector. A flask of water (B) acts as both condenser lens and heat filter. A bit of carbon (D) absorbs bubbles and prevents "rain" on the screen. 5. Turn-of-the-century projectionists hand-cranked the projector and rewound film manually. 6. Jenkins' American Phantascope projector.

5

6

early 1890s furthered the development of cinematography: Le Prince, Acres, and Paul in England, Demeny in France (Demeny had been Professor Marey's assistant and used some of his ideas which he then patented after making small improvements), and Jenkins and LeRoy in the United States.

But it was Thomas A. Edison's invention factory that did as much as anyone to make movies practical. One of Edison's bright young men, William Kennedy Laurie Dickson, a Scotsman, was given the job of finding a method of taking and showing moving pictures. In 1892 he built a camera which took forty-six pictures a second on 35mm celluloid film. The pictures had a frame size of eighteen by twenty-four millimeters, and there were four sprocket holes on each side of each frame—exactly the size used ever since, not only in movie cameras, but also in 35mm still cameras, most of which use two frames for each picture. This film configuration was Edison's big contribution.

But Edison and Dickson did not devise a projector. They built a sort of cinematic peep-show cabinet called a "Kinetoscope." This held some fifty feet of film in an endless loop. When a penny-arcade patron put his penny into the slot of the Kinetoscope, an electric motor started the film moving and also started a revolving shutter which allowed a light bulb to momentarily illuminate each frame as it went by the magnifying eyepiece. Unlike a projector, the film did not stop momentarily at each frame.

When the Lumière brothers built their "Cinematographe" they followed Edison in their use of 35mm film. But, oddly, instead of using four squarish sprocket holes on each side of each frame, the Lumières punched a single round hole at each side of the film.

Within ten years of the Lumières' first public picture show, the world was well on its way to going movie crazy. But there were still mechanical problems. Films tore and jumped because perforations were not standardized. And it took years to eliminate "flicker" (the name still hangs on as "the flicks"). But more serious was the danger of fire. At first the film that went through the projector merely ended up in a bin under the projector table. There was no means of automatically winding it up on another reel. And the film was flammable celluloid (nitrate). If the heat from the lantern set fire to the film in the film gate, the whole works burst into flame with a bang. There were some frightful theater fires because of this.

Now, of course, movies are at least mechanically perfect, but would old Professor Marey bother to point his Chronophotographe at some of the "animal movement" we see on the screen today?

7

7 The Little Servants

To Make Life Easier

There would be nothing unusual today about operating a sewing machine or a typewriter while listening to a phonograph in a room lit by an electric light. We accept these devices as perfectly normal appurtenances of our daily lives. Yet, not much over a century ago, none of them existed.

A mid-nineteenth-century woman laboriously sewing a shirt for her husband would be straining her eyes under an oil lamp. And perhaps, instead of a symphony from the speakers of a phonograph, she might be listening to the scratching of a pen as her husband wrote a letter asking for a job in a buggy-whip factory.

The phonograph today is still primarily an instrument of pleasure. But without the others—the typewriter, the sewing machine, and the light bulb—our present-day economy could hardly exist.

Consider the typewriter. It is impossible to imagine a modern business surviving while depending on scriveners laboriously writing out in longhand the company's correspondence, orders, forms, and whatnot. Imagine the huge armies of ink-stained clerks General Motors would need for its paperwork.

In addition to revolutionizing business, the typewriter contributed to another major advance of the last hundred years: the emancipation of women. Until the typewriter came along, ladies didn't work in offices. The manual writing-out of business communications was done by male clerks, who wore eyeshades and cuff protectors while standing at high slanted desks.

But when the first Remington appeared, for some reason it was immediately assumed that typewriters would be operated by females. As early as August, 1872, *Scientific American,* when describing the new writing machine, referred to the operator as "she."

Previously, poor young women hadn't a prayer of getting jobs in business offices. They either became domestics or worked in factories, usually in textile mills or as seamstresses. Certainly many of them who became secretaries were merely substituting one kind of manual labor for another—they pushed a typewriter instead of a broom. But the bright ones showed that they could do more than merely push typewriter keys—and the woman executive was born.

The sewing machine had an equally important effect on the economy. Before it came into use there were few mass-produced suits or dresses or shirts. Machine-made shoes were unknown. Only the rich could afford to wear properly made clothes, hand-stitched by tailors or seamstresses. The poor wore ill-fitting clothes, usually made by women working at home. Their shoes were pegged together by cobblers, and must have been excruciatingly uncomfortable—it wasn't until they were made in quantity by machine that there was any difference between a right and a left shoe.

The sewing machine made modern clothing factories possible, and with the factories came hundreds of thousands of jobs. They were not always good jobs, for until labor unions forced reforms such factories were often twelve-hour-a-day sweatshops.

But most sewing machines did not end up in factories. Due to installment selling, which the Singer Company innovated in 1856, the sewing machine became, long before the vacuum cleaner and the washing machine, the most common of household appliances. Today there are more than 40 million of them in use in the United States, and many millions more elsewhere.

Important as they are to our civilization, the typewriter and the sewing machine are less necessary to our lives than the electric light. True, until well into the 1890s, fairly tall buildings, steamships, railway cars, and other such modern places and means of transport were lit by gas or oil. Until 1912 automobiles burned acetylene gas in their headlights. But can you imagine an eighty-story skyscraper or the rushing masses of motor cars on a turnpike depending on gaslight?

The phonograph is of perhaps lesser importance. We think of it primarily as a means of entertainment (except when our offspring turn it on in full stereophonic intensity to assail our ears with the peculiar sounds they claim to love).

Preceding pages: Phantom electric light bulb dominates this painting of Edison's Menlo Park laboratories. Opposite: Daguerreotype of Elias Howe. Left: French tailor Barthelemy Thimmonier patented the first practical sewing machine in 1830. This is a copy of his original machine.

But the phonograph does more than merely entertain. It educates in the appreciation of good music. Further, the phonograph, like the sewing machine, the typewriter, and the electric light, has created a new and lucrative industry which has made some canny types very rich indeed.

Stitches by the Sextillion

The most common mechanical contrivance on earth is the sewing machine, the ubiquitous assembler of mankind's clothing. Japanese kimonos, Indian dhotis, Masai robes, Tibetan jackets, South Sea sarongs, Savile Row lounge suits, every kind of coat, pants, shirt, blouse, boot, or cap worn by every race of man, woman, or child (except, perhaps, the handmade fur garments of the Eskimos) all have had their parts joined together by a sewing machine.

Even the great Indian leader, Mohandas K. Gandhi, who warned his followers against western machinery, learned to use a sewing machine while in prison and called it "one of the few useful things ever invented."

Ask any schoolchild the inventor's name and he'll say, "Elias Howe," as automatically as he couples Fulton with the steamboat and Morse with the telegraph. But long before Howe there were ingenious fellows who worked out methods for mechanically pushing a needle and thread through cloth or leather and leaving in their wake an even line of stitches. The first of them appeared in 1790, during the Industrial Revolution. He was Thomas Saint, a cabinetmaker of Greenhill Rents in the parish of St. Sepulchre, England, who patented the first sewing machine we know of.

Saint designed his machine to stitch leather boots and shoes. Mostly it was made of wood, but its working parts were metal. The leather was placed on a traveling carriage under a forked needle which moved vertically from an overhanging arm, while a looping device worked from underneath to form a chain stitch. Although Saint's machine had important features which are essential to sewing machines today, his patent drawings lay forgotten for many years because the patent office filed them with those concerning glues for leather. In any case, Saint for some reason never seems to have built more than one experimental machine. Nor did anyone find a practical application for it.

During the next few decades other inventors built sewing machines, some of which might well have been developed. But, despite the obvious fact that such a machine was desperately needed by the rapidly expanding populations of Western Europe and America, there was savage opposition from the many thousands of factory seamstresses and tailors who were earning meager livings sewing by hand in the recently organized ready-to-wear clothing industry.

Still, it was a poor French tailor, Barthelemy Thimmonier, a man colossally ignorant of anything mechanical, who first had some success with a machine for sewing seams for garments. In 1825, obsessed with the idea, he neglected his near-starving family and his small tailor shop to build his prototype. Four years later he had a crude but workable machine. Built mostly of wood, it used a barbed needle (like a crochet needle) to make a chain stitch in which the loops lay on the upper surface of the cloth. The French government granted a patent in 1830, and Thimmonier put his device to work making uniforms for the French army. The following year, however, a screaming mob of tailors broke into the factory and smashed the machines—eighty in all—to kindling wood. Thimmonier was lucky to escape alive.

Out of business, he scratched a living peddling hand-built machines for fifty francs apiece. By 1848 he was on his feet again with a much-improved machine capable of two hundred stitches a minute on almost any kind of material, from muslin to leather. But again a mob raided his premises and destroyed his machines.

He was able to salvage and repair one machine, which he took to England and patented there. In 1850 he patented it in the United States, as well. But by then he had been outdistanced by other inventors with better machines. Although his model was the first one practical enough to produce in quantity, he never made money from it and died

Howe sewing machine of 1845. Howe's brother sold the rights to this machine to William Thomas in England. Opposite, left: 1874 model of Thomas Saint's 1790 sewing machine, shown at the Paris exposition of 1878. Right: First Singer sewing machine, patented in 1851.

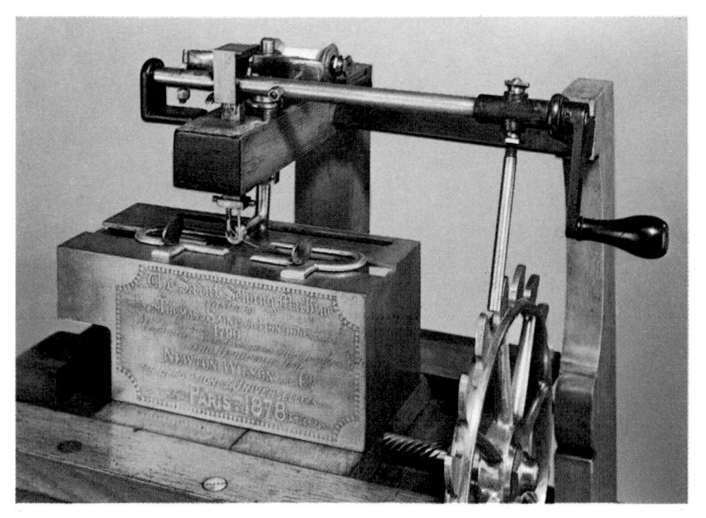

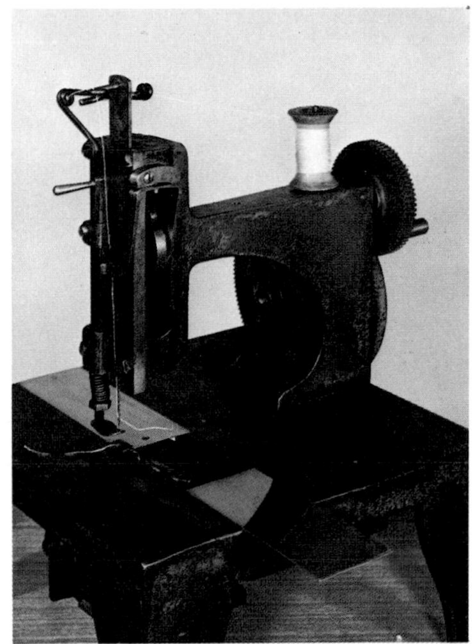

broke in 1857.

During the early 1830s, while Thimmonier was struggling to perfect his model, a prodigiously gifted American inventor, Walter Hunt, became intrigued by the sewing machine. Inventions were easy for Hunt. Without apparent effort, ideas popped from his head in a shower. He invented the safety pin in three hours one afternoon (and sold the rights for $100). He also invented the paper collar, a nail-making machine, a repeating rifle, a snow plow, and a fountain pen, among innumerable other things.

Hunt called the device he built in his shop on Amos Street in New York a machine "for sewing, stitching and seaming cloth." Its most important feature was a needle with an eye near its point. Moved through two pieces of cloth it left a loop of thread on the other side of the cloth. At the same time a shuttle inserted a second thread into the loop which was then tightened by a reverse movement of the needle, thus forming a locked stitch which firmly resisted unraveling. Although it made stitches as strong as those by present-day machines, Hunt's could only make straight seams; the work couldn't be turned.

Hunt enthusiastically encouraged his young daughter Caroline to go into the business of corset manufacture, using his machine. But she, like many others in the 1830s, had strong views against mechanization. She convinced Hunt that his machine would cause unemployment and starvation. He gave in and turned to other inventions without patenting his quite advanced model. He had second thoughts some twenty years later, in 1854. But when he applied for a patent he was turned down on the grounds that he had abandoned his invention.

During the 1830s and 1840s patent offices in almost every country were deluged with ideas for sewing machines. None of them was successful, at least partly because their inventors weren't persistent or tough enough. But a hard-headed little man barely over five feet tall, named Elias Howe, Jr., was tough and persistent and eager to get rich. Howe was twenty-one, very poor, and working as an apprentice in a Boston machine shop which built inventors' models when he became imbued with the idea of inventing a sewing machine. His interest was sparked by an argument between his boss, Ari Davis, and a well-dressed customer who was interested in building a knitting machine.

"Why are you bothering with a knitting machine?" Davis bellowed. "Why don't you make a sewing machine?"

"Can't be done," the customer answered.

"I can make a sewing machine," said Davis.

"You do that, Davis," said the customer, "and I'll assure you an independent fortune."

The words "independent fortune" excited young Howe, who was having a very rough time supporting a wife and three children on the $9 a week Davis was paying him. Davis soon forgot his rash claim, but Howe remembered.

Howe overheard the argument between his boss and the elegant stranger in 1839, the same year that Hunt put his machine aside. But it wasn't until four years later that Howe actually tried to build a sewing machine. Working at night and with no knowledge of what other inventors had done, he first tried to mechanically simulate the actions of his wife's arm and hands when sewing. His needle was pointed at both ends and had its eye in the middle. He soon gave up that notion as unworkable. Next he came up with the idea of using two needles, one curved and with an eye near its point, moving sidewise through the cloth, the other on the other side of the work, like a shuttle through the loop made by the horizontal needle. This, of course, was much like the method Walter Hunt had used, but there is no evidence that Howe ever heard of Hunt's device.

In his enthusiasm for the machine's potential, Howe quit his job and found a partner, an old acquaintance named Fisher, who was willing to take Howe's family into his home and feed them while Howe set up a workshop in his attic. He also put up $500 for tools and materials. In return he was to get an equal partnership in the invention.

Working frantically to perfect his machine, Howe was, by mid-1845, able to use it to sew up two woollen suits. He then arranged a competition at the Quincy Hall Clothing Manufacturing Company in Boston. Howe and his machine sewed five seams. Five Quincy seamstresses each sewed one seam. Howe not only finished first, but his five seams were

Sewing machine makes proliferated during the nineteenth century.
1. Grover and Baker machine of 1871.
2. 1861 Wheeler-Wilson machine.
3. Ornately cast 1872 Weir sewing machine.
4. Thimmonier and his 1830 machine. 5. Willcox and Gibbs model, patented in 1857.
6. The last machine bearing Howe's name and medallion portrait was made in 1888 by Howe's sons-in-law. Howe died in 1867.

2

1

3

4

5

6

judged better than the work of the five women.

Howe got his patent in September, 1846. But his machine still had serious faults. It did not automatically move the material along. Instead, the cloth to be sewn was suspended by pins from the edge of a thin metal rib called a baster plate. If a seam was longer than six inches, the machine had to be stopped while the fabric was moved along by hand and rehung on the baster plate. Nor could it sew a curved seam.

Howe found it impossible to sell his machine in the United States. Tailors and seamstresses derided it as impractical. Fisher, who had invested about $2,000, stopped his contributions.

Howe thought the machine might receive a better reception abroad and sent his brother Amasa to show it to English businessmen. One manufacturer, William Thomas, who made corsets, umbrellas, shoes, and other items, bought the English rights. He paid £300 ($1,500) and contracted to pay a royalty on each machine he sold, except for those he used in his own factory. As part of the deal Howe was to come to England to devise a special version of the machine to sew leather.

Howe, his wife, and their three children duly arrived in England. The results were disastrous. Thomas didn't honor his commitment. He treated Howe as just another of his wretched factory workers and failed to pay the promised royalties. Howe left Thomas, but, nearly destitute, he was unable to hire a barrister to sue him. He had to borrow in order to pay his family's passage home. He soon followed, first pawning his machine and patent papers to buy a ticket in steerage. To feed himself he worked as a cook for the immigrants bound for New York.

In New York early in 1849 he tried to get a job as a machinist. His wife lay dying of consumption in Cambridge, and he had to borrow $10 to make a trip to her bedside. At the time of his wife's death Howe was only thirty-three, but disappointment and disaster made him a silent and bitter man, old beyond his years.

During the few years Howe had spent in England, the sewing machine had achieved a modicum of popularity in the United States. But no one seemed to remember his work and his patent, and other men had made important improvements and patented them. John Bachelder, for example, had patented a machine which fed the material continuously above a horizontal work plate. It and a number of other machines were doing work in factories when Howe, outraged, decided to sue for infringement of his basic patent. But first he had to borrow enough money to redeem the machine and his patent papers from the London pawnshop. Further, he was determined to go after Thomas, who was manufacturing his machine in England without paying royalties. He was angrier at Thomas because he felt that if had he received his rightful royalties, the money would have saved his wife's life.

But the ensuing court battles, paid for in part by a George Bliss, who had bought ex-partner Fisher's share, were but the beginning of what became known as the "sewing-machine war." And the biggest adversary in that war was the man whose name has been on hundreds of millions of sewing machines since: Isaac Merritt Singer.

Singer was born in Oswego, New York, of poor German immigrant parents. Almost illiterate, he left home when he was twelve and tried to be a farmer, an actor in an itinerant Shakespearean troupe, a machinist, and an inventor of machines for excavating rock and carving wood. By the time he became interested in the sewing machine he was a first-class machinist and had become a pretty tough customer.

Singer first saw a sewing machine in the shop of Orson C. Phelps in Boston. Phelps was building crude machines for one of the numerous infringers of Howe's patent, but was going crazy trying to make them work reliably.

Singer, who was in Boston to promote a carving invention, carefully examined the machine and suggested an improvement to the shuttle mechanism. Then he came up with an idea that would make the sewing machine truly practical. The machine he conceived was basically the one which would become a world-wide success. "In place of the needle bar pushing a curved needle horizontally," he said, "I would have a straight needle and make it work up and down."

1

2

3

4

5

For over a century the sewing machine has been one of the most widely used devices on earth. 1. In nineteenth-century Pennsylvania. 2. In the Caroline Islands, about 1900. 3. By a Berber woman in Morocco. 4. At 12,500 feet, in the Peruvian Andes. 5. By operators in a New York sweatshop, about 1912. 6. By a Ceylonese couple, 1892.

CEYLON
COPYRIGHT 1892, BY THE SINGER MANUFACTURING CO.

6

Phelps was a skeptical Yankee. "Draw me a sketch of your plan and we'll look at it," he said.

Overnight, Singer made a drawing and showed it to Phelps and his assistant, Zieber. In addition to his straight needle with an eye near its point, his sketch showed an overhanging arm, a table to support the cloth, a presser foot to hold the cloth down and a roughened wheel sticking up through a slot in the table to move the material along. Gears would work the needle arm and the shuttle. And the machine could make straight and curved seams of any length.

Later Singer said, "Phelps and Zieber were satisfied that it would work. I had no money. Zieber offered forty dollars to build a model machine. Phelps offered his best endeavors to carry out my plan and make the model in his shop. If successful we were to share equally. I worked at it day and night.

"The machine was completed in eleven days. About nine o'clock in the evening we tried it; it did not sew. Sick at heart, about midnight, we started for our hotel. On the way Zieber mentioned that the loose loops of thread were on the underside of the cloth. It flashed upon me that we had forgotten to adjust the tension on the needle thread. We went back, adjusted the tension, sewed five stitches perfectly, and the thread snapped, but that was enough. I took it to New York to patent it." Singer got his patent in 1851.

All the other sewing machines then in use were powered by means of hand cranks. Singer, who had been using the machine's packing box as a sewing table, got the idea of using a treadle like those used to turn spinning wheels. For a couple of years he didn't realize that the treadle was patentable. When he was at last advised of this by a competing manufacturer, he found that it was too late under patent law, since he'd used the treadle for two years.

Singer was not only a fine machinist, he was also a first-class promoter. He showed his machine at county fairs, church suppers, and other such mid-nineteenth-century conclaves. But mostly he impressed garment manufacturers with the excellence of his sewing machines. He set up a fine new factory and installed himself in fancy offices.

Then Howe struck. He demanded $25,000 for the infringement of his patent. Singer was doing well, but he did not have that much loose cash. He let Howe sue. The suit went on for some five years, and in the end Howe won because Singer had been using Howe's basic invention—the needle with the eye near its point.

While the suit was in progress, other sewing machines, by other inventors, reached the market. Howe sued them, too, and soon was collecting a $25 royalty on every machine. But this handsome income didn't satisfy him. He longed to see his name on a machine. He thereupon started to manufacture a machine which infringed on the patents of Singer and the others. Now they sued him.

Singer ended this "sewing-machine war" by proposing a patent pool, one of the first ever. The manufacturers involved agreed to pay Howe $15 per machine. This was reduced to $7 in 1860.

Howe at last got his "independent fortune"—and then some. He was reputed to have had an income of $4,000 a day in the 1860s when factories frantically bought machines to sew the uniforms of the Civil War. He was wealthy enough to personally underwrite the cost of a Union regiment—in which, incidentally, he served as a private.

Isaac Singer became a multimillionaire. The Englishman, Thomas, who'd bought Howe's English patent for £300, became a millionaire. And Howe's brother Amasa became a millionaire.

Poor old Thimmonier did not become a millionaire.

The Keyboard Kills the Quill

The typewriter, a notable, early eighteenth-century idea, is a prime example of the accelerating rate at which inventions occur once a precedent is established.

In Europe it was not until the fifteenth century that movable type and the printing press offered an alternative to

New-York March, 13, 1830

Dear Companion,

I have but jest got my second machine into opperation and this is the first specimen I send you except a few lines I printed to regulate the machine, I am in good health but am in fear these lines will not find you so andthe children from the malencholley account your letter gave me of sickness and deaths in our neighbourhood, I had rested contented to what I should if it had been summer season adout the health of my family, as it is jenerlly healthy during the winter months; but their has ben an unusual quentity of sickness heare this winter, and it has ben very cold in Urope as well as in America, a strong indication of the change of seasonth that I have so often mentioned.— Mr Sheldon arrived here four days ago he went imediately on to Washington and took my moddle for the Pattent Office, he will returne here next week at which time I shall put my machine on sale and shall sell out the pattent as soon as I can and return home, at aney rate I seall returne home as soon as the Lake nevigation is open if life and health is spared me. I have got along but slow since I have been here for the want of cash to hire such help as I wanted; I have been as prudent as I could, have taken my board with a family from Cyuga who keep a bording house they are very good christian people and are kind to me. I pay three Dollars a week for my board.—You must excuse mistakes, the above is printed among a croud of people asking me maney questions about the machine. Tell the boys that I have some presents for them. If I had aney news to communicate I would print more but as I have none I must close hopeing these lines will find you well. I wish you to write as soon you receive this, do not make aney excuses I shall like see it in aney shape

William A. Burt.

the laborious handwriting of manuscripts and documents. This was some eight hundred years after the Chinese evidently made the same discovery, and Lord knows how many millennia after mankind first learned to incise letters or characters on clay or stone. But once the speed, uniformity, and multiple copies of printing were appreciated, men's minds turned to writing machines—something that would perform the chore of penmanship without requiring the equipment of a print shop. This great typographical leap forward took place a mere two hundred years or so later. And all the high-speed, self-correcting, do-anything, electrified machines of today have come upon us less than a hundred years after the perfected manual typewriter was hired on to do the world's business (with file-cabinetmakers and storage-warehouse proprietors as special beneficiaries).

Probably the first man's mind turned to writing machines was that of Henry Mill, an English engineer, who got a Royal Letters Patent from Queen Anne in 1714. His device was described as: "An Artificial Machine or Method for the impressing or transcribing of Letters singly or progressively one after another, as in Writing, whereby all writings whatsoever may be engrossed in Paper or Parchment so neat and exact as not to be distinguished from Print; that the said Machine or Method may be of great use in Settlements and Publick Records, the impression being deeper and more lasting than any other Writing, and not to be erased or counterfeited without Manifest Discovery."

No one seems to have found any drawings of Mills's "Artificial Machine." Or a model. If he ever built one, it has disappeared.

For the next century, the writing machines we know about generally were devices for the blind that embossed letters or symbols to be read by the fingertips. The most successful of the systems was that devised by Louis Braille in 1829. One exception was a typewriter built in 1823 by an Italian, Pietro Conti of Cilavenga. He claimed that it "wrote fast and clear enough for anyone, even those with poor sight." It was demonstrated before the Académie Française, and the savants bought the idea for possible development at a price of 600 francs.

The first American typewriter was invented by a thirty-seven-year-old Massachusetts Yankee named William Austin Burt. Like many another young American, he headed west after serving in the War of 1812 and settled his family in the vicinity of Detroit. He was a sort of backwoods mechanical genius, a surveyor, a builder of flour mills and sawmills. In 1826 he was elected to the Territorial Legislature and soon found himself drowning in political paperwork—legislative reports, bills, letters, records.

Naturally, he looked for a mechanical solution to his pen-and-ink problems. He borrowed some type from the composing room of the Michigan *Gazette*, which was edited by a friend, John P. Sheldon, and in a lean-to behind his cabin started to build a typewriter. He called it a "Typographer." It was built inside a big wooden box and was mechanically similar to the toy typewriters children still play with in which rubber type is arranged around a movable circle.

Burt's machine used half a circle and his type was metal. The semicircle was rotated until the desired letter came to the spot where it was to be printed. It could then be pressed against the paper by means of a lever. The type was inked by a pair of inking pads. The Typographer wrote on a roll of paper from which page-size lengths could be torn. An indicator on the front of the box measured the paper into letter-sized pieces. The Typographer wrote neatly and clearly. But even Burt recognized his machine's most serious shortcoming: A man with a pen could write faster.

When Burt's friend Sheldon saw the Typographer he became wildly enthusiastic and banged off a letter to Andrew Jackson, the President of the United States.

SIR:

THIS IS A SPECIMEN OF THE PRINTING DONE BY ME ON MR. BURT'S TYPOGRAPHER. YOU WILL OBSERVE SOME INACCURACIES IN THE SITUATION OF THE LETTERS: THESE ARE OWING TO THE IMPERFECTIONS OF THE MACHINE, IT HAVING BEEN MADE IN THE WOODS OF MICHIGAN WHERE NO PROPER TOOLS COULD BE OBTAINED BY THE INVENTOR. . . . I AM

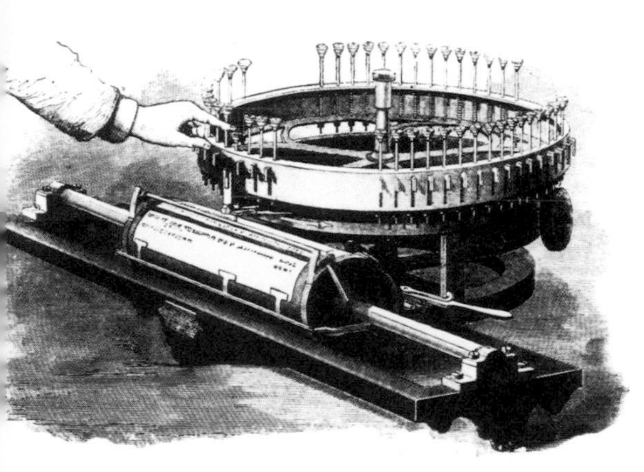

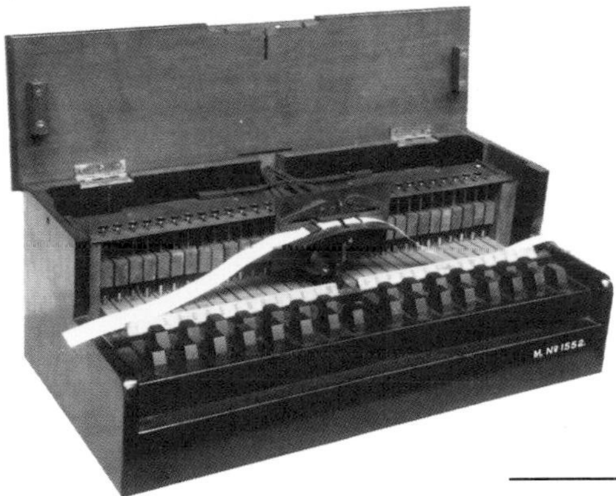

Opposite: William Austin Burt's 1829 typewriter and a letter to his wife that he typed on his machine. Middle: Charles Thurber's "Chirographer" of 1843, which used a vertical plunger for each letter. Left: Sir Charles Wheatstone's 1851 machine was designed to transcribe telegraph messages on tape.

SATISFIED . . . THAT THE TYPOGRAPHER WILL BE RANKED WITH THE MOST NOVEL, USEFUL AND PLEASING INVENTIONS OF THIS AGE.

On the back of the letter Burt typed out a formal request for a patent. And on July 23, 1829, Burt got his patent, duly signed by Andrew Jackson and Martin Van Buren.

Despite Sheldon's euphoric dreams that the Typographer would be a bonanza, no one in the rough, backwoods environs of Detroit wanted a typewriter. Nor did Burt and Sheldon have any better luck when they went to New York to raise capital.

Soon after Burt gave up on his Typographer, scores of other inventors, most of whom could not have known of his obscure device, came up with a gaggle of writing machines, some of them almost practical, others very odd, indeed. The time obviously was getting ripe for such an apparatus and, as has happened with many other inventions, awareness of the need for mechanical writing was seizing many men in many places simultaneously (even as steam locomotives were the things to invent in the early 1800s and automobiles in the late 1800s).

In 1833 Xavier Progin, a printer from Marseilles, invented what he called a "Machine Kryptographique," which had its typebars set in a semicircle, so that each letter could be struck to the same printing point (like most machines today). But instead of moving the paper to type a line Progin had his whole mechanism move sideways.

The first man to use a typewriter ribbon was Giuseppe Ravizza, a lawyer of Novara, Italy. Ravizza started to work on writing machines in 1830, when he was nineteen. Giuseppe Ravizza, a lawyer of Novara, Italy. Ravizza started to work on writing machines in 1830, when he was nineteen. He finished his first one—which he called the Cémbalo Scrivano—seven years later, and he kept on improving his models until 1885, when he died. At the end he was working on a machine which would imprint syllables.

The first machine to use a cylindrical platen, and which provided letter and line spacing, was the "Chirographer" of another Massachusetts man, Charles Thurber, which appeared in 1843. Despite the great importance of these elements, the Chirographer was otherwise archaic. It had no keyboard as we know it. A group of plungers, each of which carried a metal letter at its lower end, projected through holes in a wheel-like metal ring. You turned the wheel until the plunger bearing the letter you wanted was over the paper. Then you pushed the plunger straight down. Very slow!

Another type of plunger-operated machine was invented by Pastor R. J. Malling Hansen of Copenhagen. Called the "Writing Ball," it dispensed with Thurber's revolvable wheel. Instead, its plungers—no less than fifty-two of them—radiated from a hemisphere like so many porcupine quills. Each lettered plunger pointed toward the spot on the paper which was to be imprinted. The paper lay flat on a tiny car which ran on tracks. The "writing ball" is said to be the first machine manufactured and sold to the public.

By the late 1860s, patent-office files bulged with writing-machine inventions. One of them, the Pterotype (winged type), used a type-wheel, not too dissimilar from Burt's half-wheel device of 1829. The inventor, John Pratt, used three banks of keys to operate his wheel, however, instead of the primitive lever Burt had depended upon. Although the Pterotype was no great shakes, it was admired by Alfred E. Beach, the editor of *Scientific American* and himself the inventor of a huge machine that wrote on a paper tape. Beach reprinted a story about the Pterotype he had read in the London magazine *Engineering* (the machine had been patented in England, as well as in the United States), and a man named Christopher Latham Sholes read it. Although half a hundred others had marched down the same road ahead of him, Sholes won the world's accolade as inventor of the first commercially practical typewriter. He was Collector of Customs in Milwaukee, but had been a newspaper editor and knew something about printing. He also had been a member of the Wisconsin legislature and postmaster in Milwaukee. He was a tall, skinny fellow with long hair and a beard. At forty-eight he was beginning to stoop a bit—and no wonder. He had a wife, ten children, and a grandchild to worry about. Sholes was a tinkerer, unremarkable for a man

1. 1864-65 Pratt typewriter. 2. Later Pratt typewriter, the "Pterotype." 3. Sholes's 1867 machine. 4. Christopher Latham Sholes. 5. Sholes's 1868 patent model typewriter. 6. 1872 Sholes Type-Writer had a keyboard similar to that used today, but its carriage was moved by a weight on a string. 7. Remington-built Sholes and Glidden typewriter.

1

2

descended from Yankees, and he hung around with other tinkerers in a machine shop run by C. F. Kleinsteuber. Sholes was struggling with a mechanism he'd dreamed up for automatically numbering the pages of books. A draftsman and engineer, S. W. Soule, was helping him. Other hangers-on in the machine shop were Carlos Glidden, who was involved with the invention of a steam-driven rotary plow, and Dr. Henry W. Roby, a court reporter who was fiddling with a peculiar clock to be used by a magician.

Things moved at a leisurely pace at Kleinsteuber's. The tinkerers talked as much as they worked. Anyhow, most of the actual labor was done by Matthias Schwalbach, a first-class mechanic and patternmaker. It was natural for this crew to be interested, one afternoon, when Sholes mentioned Pratt's Pterotype, which he'd read about in *Scientific American.* Glidden suggested that Sholes ought to try building something of the sort. After all, he pointed out, Sholes's page-numbering device was a kind of writing machine. Sholes admitted that he'd had similar thoughts. "I'm going to try it," he said.

A few days later Sholes came up with an idea. Instead of following Pratt's system which had its type on a movable circular plate, he would put each letter separately on bars or keys, like piano or telegraph keys. Soon, with the help of his mechanical cronies, he had rigged up a one-letter demonstrator made of a telegraph key which printed the letter W by striking upward through a piece of carbon paper. Carbon paper was a rarity in the 1860s, and Sholes had had to borrow a sheet from a friend, Charles E. Weller, in the Milwaukee telegraph office.

By the end of the summer, Sholes, Glidden, and Soule had built a crude working model of a typewriter with piano-like keys. The machine was built into the top of an old kitchen table. The gang at Kleinsteuber's thought it worthy of a celebration, and a newspaperman who joined the party wrote, not on the writing machine, but in pencil:

"They let the funny thing go,
And by jingo!
It prints the lingo
Of a red flamingo
A Greek or Gringo
A monk or mingo,
Great Dane or dingo."

The celebration was a bit premature. Everybody in the shop had ideas for improving the machine. Sholes had even more such ideas than the rest of them. He started to tear down the machine in order to revise it, a chore he went through again and again for years.

Sholes and his friends had that old inventor's problem: money. They needed money to build patent models, to pay patent lawyers, to set up a factory. They offered shares in the "Type-Writer" for cash. One of the many people they approached was an old newspaper acquaintance, James Densmore of Meadville, Pennsylvania. An unlikable character, Densmore was not only an ex-newspaperman, but a lawyer, a printer, a speculator, a high-pressure salesman and promoter, and an inventor. Densmore was exactly the opposite of Sholes. Where Sholes was thin and quiet and neat, Densmore was a huge, red-bearded, ugly loud-mouth who purposely dressed like a slob.

Sholes didn't really know Densmore very well, but he'd heard that Densmore had invented the first railway oil-tank car and made a lot of money out of it (which he hadn't). So Sholes wrote him one of the very earliest typed business letters inviting his participation.

Densmore wrote right back. He wanted a piece of the action. He would pay the outstanding bills ($600) and all future expenses for a quarter interest in the Type-Writer. He paid the $600 immediately. What Sholes didn't know was that the $600 was all the ready cash Densmore had. And he'd never even seen the machine.

By the time the Type-Writer was perfected and marketed, all those involved fervently wished that they'd never heard of Densmore. But without him the machine might never have succeeded. He was the one who never lost faith when it seemed the invention would never amount to anything. He loudly trumpeted its virtues—and he could be very loud, indeed. It must be added, too, that he eventually

3
4
5
6
7
M. № 1512.

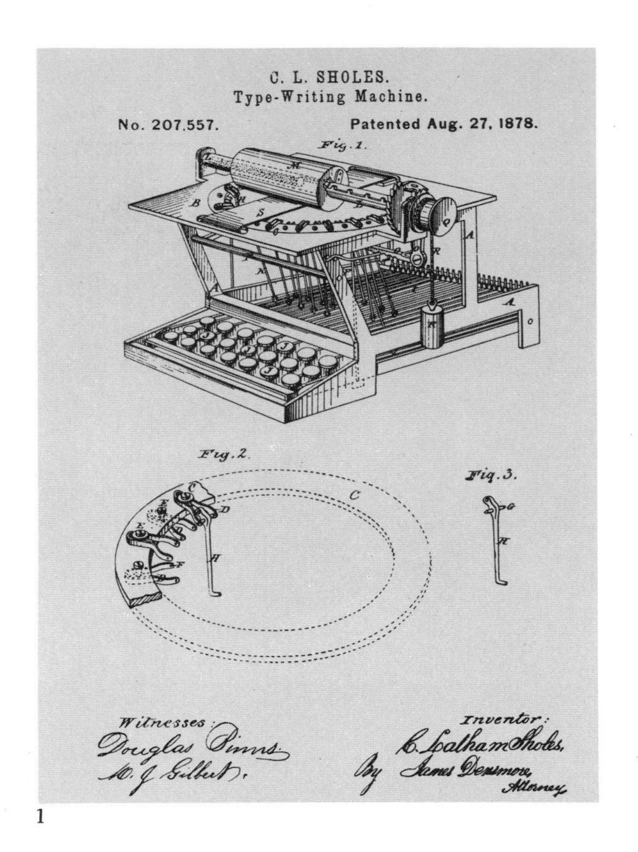
C. L. SHOLES.
Type-Writing Machine.
No. 207,557.
Patented Aug. 27, 1878.
Fig. 1.
Fig. 2.
Fig. 3.
Witnesses:
Inventor:
C. Latham Sholes,
By James Densmore,
Attorney

1

2

3

4

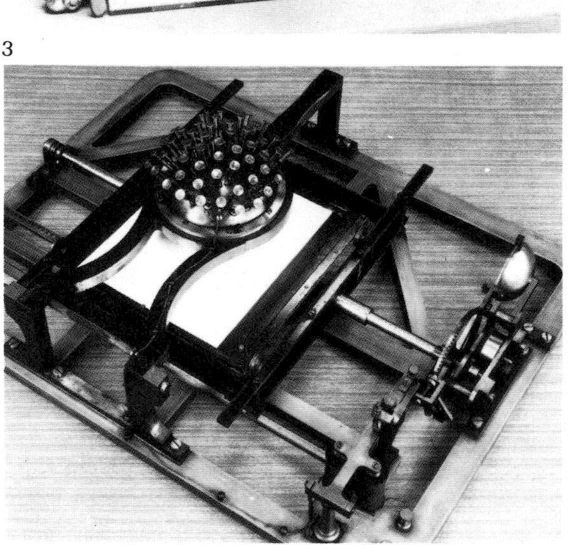

5

TYPEWRITER

6

7

1. Sholes Type-Writer of 1871 wasn't patented until 1878. 2. Beach typewriter. 3. Hammonia typewriter. 4. Velograph "index" typewriter. 5. Hansen "writing ball" machine. 6. Bar-Lock typewriter. 7. Oliver was one of many makes that became popular but eventually faded away.

grabbed off most of the rights and royalties.

For six years, until 1873, Densmore drove Sholes almost crazy. Bellowing and waving his arms, he ranted that the machine had to be much improved or it would not sell. Poor Sholes built some forty-odd Type-Writer models after the first one was patented in June, 1868.

For a time Sholes had great confidence in one of these models, the so-called "axle" machine, which used a cylindrical roller. But oddly, instead of the lines of type proceeding horizontally, on the long axis of the roller, they ran in a circle *around* it.

Luckily, for generations of typists to come, who might have developed permanent cricks in their necks from looking at vertical lines of typing, the "axle" machine soon evolved into later models whose typed lines ran in a more convenient direction. It was quite a while, however (long after the Sholes-Glidden-Soule machine was successfully commercialized), before the typist could see what he was typing, since the keys struck from beneath the roller.

As new experimental machines were devised, Densmore persuaded professional stenographers—Sholes's friend Weller was one of them—to borrow Type-Writers. They were asked to pound away at them without mercy. Naturally the crude machines broke down. The typebars jammed and tangled each other. The ribbon snarled and, being hand-inked, smeared both the mechanism and the typists. The carriage wasn't moved sideways by a spring as in later machines, but by a weight which worked an escapement by means of string. The string sometimes broke and the heavy weight crashed onto the table—if it didn't end up on the typist's toes.

In the autumn of 1870 an attempt was made to sell the invention to the Automatic Telegraph Company for $50,000. Densmore went to New York armed with powers of attorney from Glidden and Soule. There he met Sholes, who had lugged one of the heavy machines on the train from Milwaukee. The officials of the company had with them a serious, as yet unknown, young expert, Thomas A. Edison, whom they had hired to improve their automatic telegraph.

After talking to Edison, the company officials decided to hold off for a while. They took the model with them to examine more carefully.

After tense weeks of waiting, the Automatic Telegraph bigwigs said no. Their young man Edison could build a much better machine for $50,000. They'd wait and see. Edison did finally design and patent a "type-writing" machine. But it wasn't a typewriter. It was an electrically operated machine which later evolved into the stock ticker.

Attempts were made to sell the Type-Writer to other investors, but to no avail. Densmore, although threadbare by now and reduced to living on apples and crackers in miserable hotel rooms, and borrowing to finance the machine, still had unlimited confidence in the Type-Writer. "I believe in the invention from the topmost corner of my hat to the bottom-most head of the nails of my boot heels," he bellowed. He had to believe in it. For by 1872 he owned most of it. He had raised most of the money needed to pay the bills for the patent models, the experimental machine work, the travel expenses.

Some of the early shareholders, who had given up the hope that any profit would ever be realized, had gradually turned over their shares to Densmore. Glidden, Roby, and Schwalbach owned nothing.

Densmore and Sholes, who still owned part of his invention, had several times tried to have Type-Writers manufactured in Kleinsteuber's shop, but the attempts had failed. The small, underequipped shop was inadequate for serious production. In the summer of 1872 Densmore rented an old stone building near Milwaukee. It had a water wheel for power, and Densmore equipped it with some secondhand tools. A few workmen were hired and Schwalbach was installed as their foreman.

Here much-improved new "continuous roll" machines were manufactured, one by one. *Scientific American* described the new model: "The operator is . . . before a keyboard or assemblage of knobs, each of which is marked with a letter or punctuation mark. The paper is placed on an endless belt and then passes over [under] the cylinder, situated on the top of the box enclosing the lower portions of the machine. The cylinder rests in a frame on wheels, and is

It was assumed almost immediately that the typewriter would be operated by females. Right: 1872 engraving shows a young lady typist slaving away at the keyboard. Opposite: By 1911 thousands of women were part of the formerly all-male business world.

made to move bodily in the direction of its length by means of a weight. We will now suppose that the operator begins to write. As she presses a key, it not only causes the type to fly up and leave its imprint on the paper, but, at the same time, it moves a rock shaft and dog, which acting on a rack permits the cylinder to be drawn, by the falling weight, a space equal to the proper distance between the letters in a word. The word being finished, the longer interval between it and the one following is obtained by pressing down the square frame extending beyond the keys in front. . . . As soon as the cylinder has traveled the length of a line, it strikes a bell, thus notifying the operator of the fact. By pressing down the treadle under the machine, the cylinder is drawn back to its starting point, the weight raised ready to descend again, and at the same time a lever is moved which, acting on a ratchet wheel on the side of the cylinder, causes the latter to rotate on its own axis a sufficient distance to place the paper in a position to receive the impression of another line."

It is interesting to note that even at this early date it was assumed that the operator of this machine would be a woman.

One feature that *Scientific American* did not mention was a new keyboard. Arranging the keys in alphabetical order had not worked too well. The most-used letters were not always the easiest for the fingers to reach. Furthermore, some of them were so close together that their typebars often collided. Sholes and Densmore had been printers and knew that in a type case, the pieces were arranged according to convenience, not alphabetically. They rearranged the keyboard to QWERTYUIOP etc., and that is how it has remained. The letters were still all upper case. Lower case—with a shift for capitals—came in 1878.

The Milwaukee-built machines sold quite well, although not to government departments where Densmore hoped they would generate publicity. They went to telegraphers, reporters, lawyers, and one went to Allen Pinkerton, the detective who claimed he used it "almost daily." The factory lost money. It cost more to build Type-Writers than could be got for them. Densmore, although outwardly optimistic, saw money troubles ahead.

Now another man got into the act—an old friend of Densmore's from the oil fields, George Washington Napoleon Yost. Yost was the ultimate smoothie. In spite of his beginnings as a poor farm boy from upstate New York, he dressed impeccably, spoke in cultured accents, and gave people the impression that he was of the upper crust and certainly rich. All of which he wasn't. He was a tough little promoter and, like Densmore, a product of the rough-and-tumble oil fields. He was also an inventor. He had devised a sowing and reaping machine and owned a factory which produced it.

In December, 1872, Densmore, who had by then persuaded Sholes to devise a new model with a spring-powered carriage, invited Yost to come to Milwaukee to see it. Yost spent a day at the little factory, and before leaving told Densmore that he knew of a far better one. It was in Ilion, New York, and the people who ran it were E. Remington & Sons.

The Remingtons were, of course, the rifle and pistol manufacturers who had helped arm the Union during the Civil War. Their huge plant mass-produced great quantities of arms as well as fifty Remington Empire Sewing Machines each day. Densmore and Yost lugged the latest model Type-Writer to Ilion, set it up in their room in Small's Hotel, and demonstrated it to Philo Remington, to a young executive named Henry H. Benedict, and to two of the company's top mechanics, Jefferson M. Clough and William K. Jenne.

Yost was never more eloquent than in the hard-sell spiel he delivered. The Remington people were crazy about the machine, but played cool in bargaining for it. Finally a contract was signed; Remington would not buy the invention, but would have its mechanics remodel the machine. They agreed to build a thousand of them, which Densmore and Yost would pay for. Through his own sales talks Yost had convinced himself that the Type-Writer would become a bonanza and had bought a share in it from Densmore. The firm of Densmore & Yost would market the machine.

The first Remington-built machine appeared a year later. It was encased in black enameled metal bedizened with gold swirls and flowers, and stood on a curly cast-iron stand. With its treadle for returning the carriage it looked

remarkably like a sewing machine. And no wonder. The head mechanic of Remington's sewing-machine department had had a big hand in its redesign. There was one thing Densmore didn't like about the machine—its name. It was emblazoned "Sholes & Glidden Type-Writer." He felt it should have been Sholes & Densmore. But Glidden had managed to get back into the act by suggesting improvements and demanding a share in the company on the strength of that contribution.

The rest of the story of the typewriter is much like that of other inventions: Troubles with marketing, corporate shenanigans, rival inventors, recriminations—especially on the part of Sholes, who felt he had made too little money out of the invention. No one really knows how much he made. Guesses vary between $12,000 and $40,000. Densmore and his heirs got perhaps ten times as much. But the real gainers were all those millions of people who had for so long slaved with pen and ink.

Yet we must quote one ungrateful gentleman, Mark Twain. Walking down a Boston street with another humorist, Petroleum V. Nasby, he saw a Model I Remington in a typewriter-store window and bought it. He later wrote to the company:

GENTLEMEN:

PLEASE DO NOT USE MY NAME IN ANY WAY. PLEASE DO NOT EVEN DIVULGE THE FACT THAT I OWN A MACHINE. I HAVE ENTIRELY STOPPED USING THE TYPE-WRITER FOR THE REASON THAT I NEVER COULD WRITE A LETTER WITH IT TO ANYBODY WITHOUT RECEIVING A REQUEST BY RETURN MAIL THAT I WOULD NOT ONLY DESCRIBE THE MACHINE BUT STATE WHAT PROGRESS I HAD MADE IN THE USE OF IT, ETC. ETC. I DON'T LIKE TO WRITE LETTERS AND SO I DON'T WANT PEOPLE TO KNOW THAT I OWN THIS CURIOSITY BREEDING LITTLE JOKER.

YOURS TRULY,

SAML. L. CLEMENS

Switch on the Light!

The electric light did not spring brightly lit from the fertile brain of Thomas A. Edison. Rather, it was the end result of much experimentation by many men during the first three quarters of the nineteenth century.

In 1808, not long after Alessandro Volta had devised the electric battery, the noted English scientist, Sir Humphrey Davy, succeeded in lighting up the darkness by means other than a candle flame or an oil lamp. Davy, working at the Royal Institution, used what may have been the most monstrously big electric battery of all time. It consisted of no fewer than two thousand cells in porcelain troughs. The electrolytic fluid consisted of sixty parts of water plus one part each of nitric and sulfuric acid. The plates were of zinc and copper.

"When pieces of charcoal," said Sir Humphrey, "about an inch long and one-sixth of an inch in diameter, were brought near each other (within the thirtieth or fortieth part of an inch), a bright spark was produced, and more than half the volume of the charcoal became ignited to whiteness; and, by withdrawing the points from each other, a constant discharge took place through the heated air in a space equal at least to four inches, producing a most brilliant ascending arch of light, broad and conical in form, in the middle."

Davy had invented the electric arc light, which to this day is used as the illuminant to project films in movie theaters. Davy used pencil-shaped bits of ordinary charcoal, which burned away fast. Further, there was no means of adjusting the distance between the charcoals. The light didn't last very long. For the next thirty-odd years the electric arc was no more than an interesting laboratory demonstration. Batteries, moreover, were huge, underpowered, and expensive to build. Over the years their design and efficiency improved, but their metal elements in open-topped jars of acid would seem impossibly crude and dangerous today.

In 1844, Frenchman Léon Foucault substituted

1. Sir Humphrey Davy.
2. Davy made the first electric arc light in 1808. Thousands of wet-cell batteries were needed to supply the current for it.
3. The Brush arc light was the typical American electric streetlamp in the 1870s.
4. New York under arc lights in 1870-80. The lights had an unpleasant blue glare.

1

2

hard carbons for Davy's charcoal, and also worked out a means of adjusting the gap between them by hand. He powered his light with Bunsen batteries created by the man better remembered for his laboratory gas burner. The French claim that a Monsieur Deleuil took the world's first artificially lighted photographs using Foucault's arc light, but four years earlier two Yale scientists obtained "photographic impressions by galvanic light reflected from the surface of a medallion to the iodized surface of a Daguerreotype plate." They used *nine hundred* batteries and are said to have got a couple of quite good pictures of the medallion.

One indubitably French innovation did occur, however, on a foggy December evening in Paris, in 1844, when strollers in the Place de la Concorde were amazed to find that they could see quite clearly through the murk, although gas lights only a few yards away were almost invisible. They were experiencing the first use of electricity for street lighting—by means of arcs, not incandescent bulbs. The Parisians were delighted, and soon arc lights illuminated the Pont Neuf and the Arc de Triomphe, among other Paris landmarks. And French scenic artists, in those days of elaborate stage designs, soon used arc lights to create startling effects. By the 1860s blindingly glaring arc lights were used for street lighting in western Europe and the United States. New York, Cleveland, and even such a small town as San Jose, California, boasted such modernity.

The arc light was (and still is) a quite simple device. Two rods of carbon are placed in contact. Electric current passes from one to the other. Then the rods are gradually separated and a white-hot arc forms between them. But carbon arcs were totally impractical for home use. They flickered and buzzed and required constant attention, despite the fact that automatic devices soon were devised to maintain the correct gap. They also used an inordinate amount of current and their harsh, blue-white glare was unbearable to live with indoors. What was needed was a simple bright light which could be turned on and off as easily as a gas light. And without using a match.

An incandescent filament inside a glass globe exhausted of air would be the answer, and in 1841 one Frederic de Moleyns patented such a lamp in England. He proposed that a platinum wire be the filament and that it receive "a shower of plumbago [graphite] particles." A year or two later an American, J. W. Starr, of Cincinnati, also proposed a semivacuum glass globe. It was to contain a thin graphite strip held between two clamps affixed to a porcelain rod which, in turn, was to be suspended by a platinum wire sealed in the globe. Moleyns' lamp didn't work. His platinum wire fused. Starr's is said to have lit up for a short time. But both men were ahead of their time. An air pump that could create a good vacuum did not yet exist. Nor had a practical dynamo been invented. Batteries were still the only source of power. Starr tried to build a dynamo—a "magneto electric" machine—but died before he succeeded.

The electric light as we know it today is mostly the result of the work of Joseph Wilson Swan in England and Thomas Edison in the United States. Swan, a prolific inventor and experimenter, was one of the first perfecters and manufacturers of photographic dry plates. He also invented the carbon process of photographic printing. In 1860, almost twenty years before Edison became involved with the incandescent electric-light bulb, Swan built an electric lamp with a carbon filament made by packing pieces of thick paper with charcoal powder in a crucible, and then baking the crucible at a very high temperature. A very fine strip of the carbonized paper was then mounted inside an evacuated bulb. He still had to use battery power, and he didn't get much more than a red glow out of his paper filament. Later, in 1878 or 1879, Swan devised a filament made of cotton thread "parchmentized" by soaking it in sulfuric acid. The cotton thus lost its fibrous character and became smooth and transparent. The treated thread was then carbonized by being wound on carbon rods, buried in powdered charcoal in a crucible, and heated at high temperature in a furnace.

By 1880 the Sprengel air pump was available and Swan could achieve a good vacuum in his bulbs. Nor was it necessary any longer to use batteries for power. The electric generator now existed. In October Swan gave the first public exhibition of electric lighting, and the next month he got his English patent.

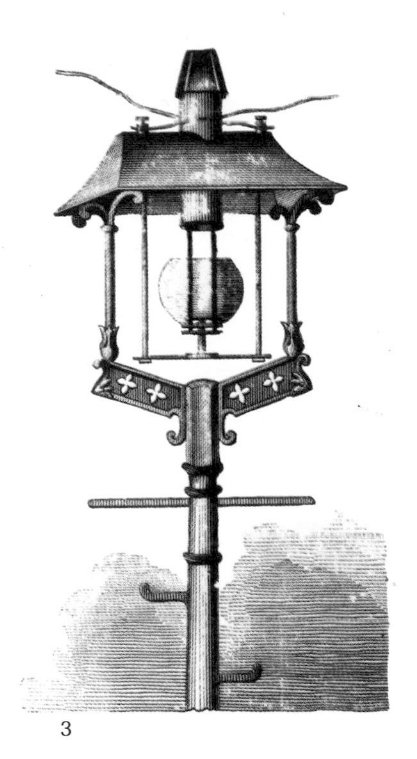
3

4

Edison first tried to produce an electric-light bulb in 1879. Like others before him he first experimented with platinum-wire filaments. But he encountered the same problems others had had. The platinum filament gave a good light when exactly the right amount of current was passed through it, but even a bit more voltage would overload and melt it. He tried various automatic devices to shut off the current when the voltage became too great. None worked; the platinum wire melted more quickly than the gadgets could react. He tried treating the platinum wire with substances he thought might make it less likely to melt: boron, silicon, and so forth. No luck. He spent a hard thirteen months experimenting before he found the stuff that would work: carbon. Strangely, he doesn't seem to have known about Swan's success with that very common element.

The story goes that Edison got the idea of using carbon almost accidentally. He was sitting at a table in his laboratory one night, mulling over his troubles with the light bulb, when he idly started fiddling with a small pile of lampblack mixed with tar. His assistants had been using the mixture during experiments with the telephone transmitter. Edison began rolling the stuff between his thumb and forefinger, thinking about other things and not paying much attention to what he was doing. After a few minutes of this, he suddenly realized that he had a thin, threadlike length of carbon between his fingers. Ready, in his travail, to try almost anything, he quickly began rolling more fine threads of the mixture. He thought he might give them a go as filaments. With one of his assistants, Charles Bachelor, he placed a lampblack thread inside a bulb and pumped out the air. (By now Edison had learned how to obtain a quite good vacuum.) The current was switched on and the filament lit up nicely. But it didn't last. Edison concluded that the failure was caused in part by air trapped in the lampblack. In addition, the lampblack thread was so fragile that it broke at the slightest jar to the bulb. But Edison knew he was on the right track.

He sent a boy out to buy a spool of cotton. He thought to try carbonized thread. A piece of the thread bent in the shape of a hairpin was clamped into a nickel mold which was then baked in a furnace for five hours. After it cooled, the mold was opened and the carbonized thread carefully lifted out. It instantly disintegrated. Bachelor and Edison tried again, and again the thread broke. They attempted it again and again and again without sleep for three days and three nights. On the night of the third day they managed to insert a carbonized-thread filament into a bulb, evacuate it and turn on the current. "It lit up," said Edison. ". . . we wanted to see how long it would burn. The problem was solved—if the filament would last. The day was—let me see—October 21, 1879. We sat and looked, and the lamp continued to burn. None of us could go to bed, and there was no sleep for any of us for forty hours. We sat and just watched it with anxiety growing into elation. It lasted about forty-five hours and then I said, 'If it will burn that number of hours now, I know I can make it burn a hundred.' "

Edison slept a whole day after his ordeal with the carbonized-cotton filament. Then he started a wild hunt for things to carbonize. Straw, paper, cardboard, wood, everything carbonizable was tried. Umbrellas and walking sticks disappeared. Nothing was safe from the furnace. Cardboard seemed to work well (as Swan had discovered twenty years earlier), but Edison was not satisfied. He took a bamboo fan and ripped off the rim. A long sliver of this made the best filament yet, but Edison surmised that there might be even better bamboo somewhere. Edison devoured everything there was to read about bamboo. He must have been taken aback when he found that more than twelve hundred varieties were known. He had to have a sample of each kind, right now. He spent some $100,000, a fortune in 1879, getting his bamboo samples. He sent men to Burma, Malaya, South China, Ceylon. He sent others to the West Indies, Guiana, Mexico, up the Amazon. The Edison laboratory at Menlo Park was inundated with bamboo. Six thousand samples were carbonized until, finally, Edison found the variety he wanted. It came from the upper reaches of the Amazon!

In January, 1880, he got his patent for "an improvement in electric lamps and in the method of manufacturing the same." In that same month the New York Board of Aldermen came out to Menlo Park to be amazed by strings of

1. Sir Joseph Wilson Swan.
2. Early experimental Swan and Edison light bulbs.
3. & 4. Later variations of Swan bulbs. Note the elaborately sprung shockproof mountings. 5. Edison's drawing of a method of mounting carbon filament in a light bulb. The date is February 13, 1880.
Opposite: The Edison screw-in socket is not yet shown in this early engraving. Nor is there a light switch. Note the gaslight-style wall fixture.

1

3

2

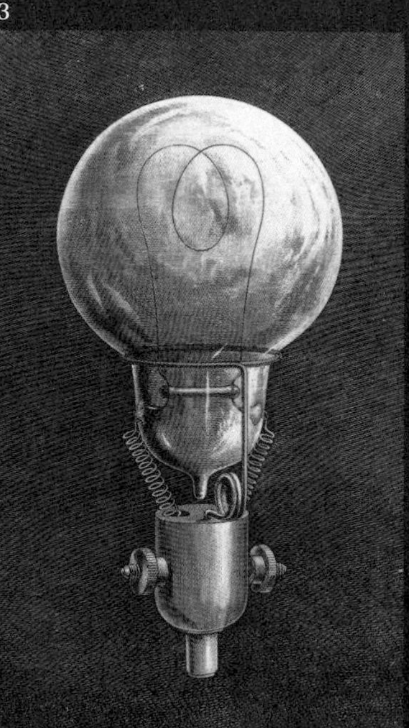

4

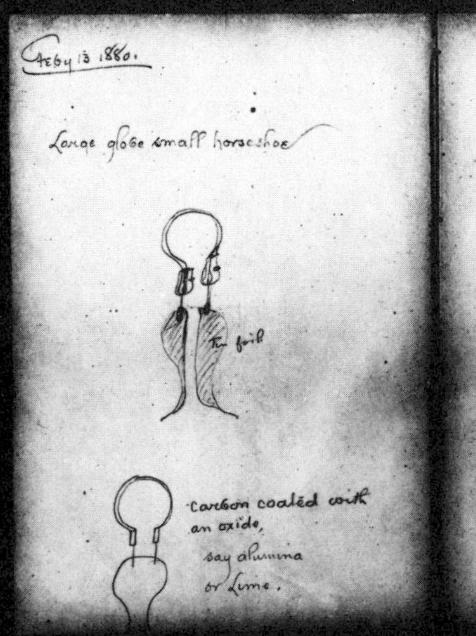

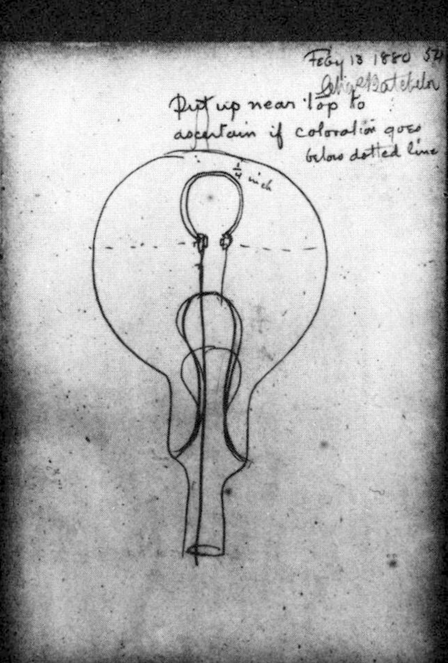

5

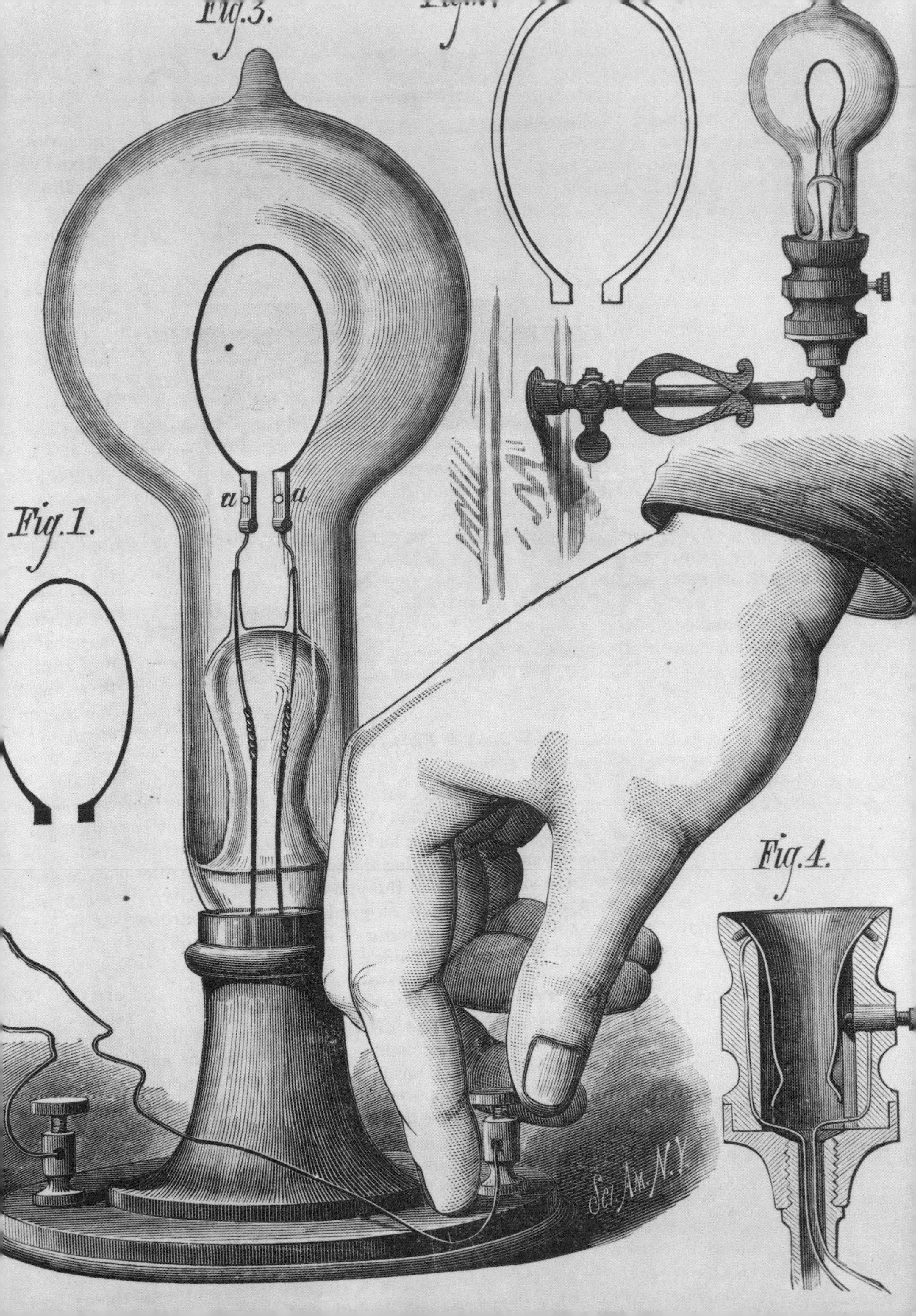
Fig. 3.
Fig. 1.
a
a
Fig. 4.
Sci. Am. N.Y.

electric lights glowing brilliantly in the winter darkness without any piping attaching them to the local gas works. Hiram Maxim, who had not yet built his giant flying machine, was one of the guests.

But electric lights are useless unless you can screw them into a socket wired to a central power station. That was Edison's next job. "I had the central station in my mind all the time that I was pursuing my investigations in electric lighting," he wrote years later in the *Electrical Review.* "I got an insurance map of New York, in which every elevator shaft and boiler and house-top and fire-wall was set down and studied it carefully. Then I laid out a district and figured out an idea of the central station to feed that part of the town from just south of Wall Street up to Canal and over from Broadway to the East River. I worked on a system, and soon knew where every hatchway and bulkhead door in the district I had marked was and what every man paid for his gas. How did I know? Simplest thing in the world. I hired a man to start in every day about two o'clock and walk around through the district noting the number of gas lights burning in the various premises; then at three o'clock he went around again and made more notes, and at four o'clock and up to every other hour to two or three o'clock in the morning. In that way it was easy enough to figure out the gas consumption of every tenant and of the whole district; other men took other sections."

Once Edison had some idea of how much current his power station might have to supply and how many customers he might have, he boldly went ahead to build America's first central powerhouse.

It wasn't easy. He had to set up his own lamp factory, his own dynamo factory. He talked a fellow named Bergmann, who had a tiny establishment on New York's East Side, into converting his shop from the making of gas fixtures to the manufacture of electric sockets and fixtures. He had to design and make switches, fuses, and meters.

Edison needed money, too. To build the station he started the first American power company: the New York Edison Illuminating Company. With $150,000 raised through the sale of stock he bought two ancient buildings on Pearl Street. While the generators and the steam engines to run them were being installed, his men started digging the first of those innumerable ditches and holes in the streets with which the Edison Company and its successor, Con Ed, have plagued New York these last ninety-five years.

On Monday, September 4, 1882, at three o'clock in the afternoon, the Pearl Street station started generating electricity. The time when man had to push back the night with an open flame was over.

The Machine That Talks Back

Who invented the talking machine? Was it Thomas A. Edison or was it that French prodigy, Charles Cros? It is true that Edison was first to have such a machine built. But the first to get the idea that it might be possible to construct the device, and also to design an excellent means of achieving it, was Cros.

Edison hadn't meant to invent a talking machine. He came upon it almost by accident. Early in 1877 he had been trying to build an instrument which would emboss the Morse code's dots and dashes telegraphically on a strip of moving waxed paper or a rotating disc. Recorded this way, the message could be repeated at a later time and at any speed. Fiddling with his telegraphic recorder, Edison tried speeding it up and was surprised to hear a musical note. It was caused by vibrations of the spring he was using to guide the embossed paper strip. This reminded him of some experiments he had made with telephone diaphragms. He got the idea that a stylus attached to such a diaphragm might record speech on a waxed tape, which could then be reproduced by drawing the tape under a stylus attached to a telephone receiver.

Edison had his shop build a rough model of such a device and then yelled, "Hallo-o-o!" into it. Running the tape under the receiver he thought he heard, albeit much distorted

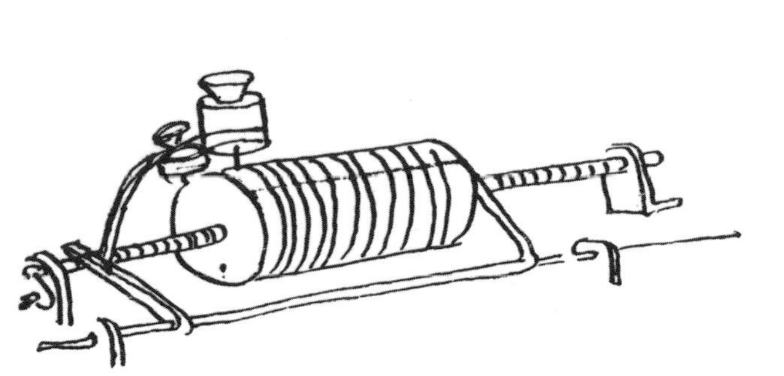

Opposite: Edison's Menlo Park laboratory in the 1880s. Note electric lighting. Middle: Edison's rough sketch of his talking machine. Left: Charles Cros, the French prodigy who invented a disc-type phonograph before Edison constructed his cylinder phonograph. But Cros' machine was never built.

and indistinctly, something that sounded almost like "Hallo-o-o!"

Edison still wasn't thinking about a talking machine. He had in mind a device which might have a connection with telephony. He thought, for instance, that someone without a telephone might make a recording of a message for later transmission from a telephone company office.

Edison seems to have been turned away from such ideas by one of his associates, a man named Johnson. Johnson had been giving a lecture in Buffalo about some of Edison's latest inventions. When he mentioned the fact that Edison had succeeded in recording the human voice on his telephone relay there was much excitement. Johnson reported this to Edison, who immediately realized that voice recording alone, not its application to telephony, was the important thing. He made a rough sketch of a brass-and-iron phonograph and handed it to one of his mechanics, John Kreusi, to build. Edison wrote "$18" on the edge of the drawing. Kreusi was paid on a piecework basis like the rest of Edison's machinists. Kreusi asked what the thing was for. "A talking machine," said Edison. And Kreusi laughed fit to bust.

Kreusi made a nice job of the little gadget. What he produced was a brass drum about four inches in diameter mounted on a threaded axle which had a small crank at one end. Turning the crank not only rotated the drum, it also moved it along the axle. A sheet of tin foil was wrapped around the drum. On one side of the drum was the recorder: a mouthpiece terminating in a diaphragm with a stylus in its center. The stylus was in contact with the tin foil. The reproducer was similar to the recorder, but was on the other side of the drum. To record, the speaker yelled into the recorder while turning the crank. The vibrations of the diaphragm caused jiggles to be impressed in the tin foil. To play back the sound, the reproducer's stylus followed the jiggles which in turn vibrated *its* diaphragm when the crank was turned. All this on December 6, 1877.

And what earthshaking phrase was the first to be so reproduced? "Mary had a little lamb."

Charles Cros, who died at forty-six in 1888, was a darling of the nineteenth-century Paris literati. He was a philologist, a doctor of medicine, a teacher of chemistry, an inventor of a color-photography process, a poet, and an accomplished comic monologist. His ideas for a phonograph were first published in *La Semaine du Clergé* in October, 1877, but he had deposited them in a sealed envelope at the French Academy of Sciences in April and asked officials there to open the envelope on December 3.

Cros' solution of the problem of making a machine talk was rather more sophisticated than Edison's. He proposed using a transparent disc with a smoke-blackened surface. The stylus attached to the voice-vibrated diaphragm would trace a jiggly spiral line showing the vibrations of the diaphragm. The spiral would then be reproduced on metal by photoengraving. To reproduce the sound a similar diaphragm, plus stylus, had but to ride the spiral groove. But Cros never had a model of his phonograph built. Although he was ahead of Edison with a better design, Edison is, perhaps rightly, considered the inventor of the talking machine.

Edison's tin-foil phonograph was no smashing success. It was far from perfect and it was expensive. It was through the efforts of Alexander Graham Bell, whose telephone had been improved by Edison, that the talking machine became much improved. Bell had received the Prix Volta from the French government in 1880. With the money that came with this honor he set up the Volta Laboratory Association in Washington, D.C. One of the projects of this electroacoustical laboratory was the improvement of the talking machine. The experiments were carried out by Bell's cousin, Chichester Bell, and Charles Sumner Tainter. Their work resulted, some five years later, in what they called the Graphophone. Like Edison's machine, it used a cylindrical record. But the recording was no longer on tin foil. It was on a wax-coated cardboard cylinder which was slid onto a rotating mandrel. The record did not move sidewise. The recorder—the diaphragm-plus-stylus—did the moving by means of a feed screw. The quality of reproduction was much better than that of the old Edison phonograph, and the canny

1

2

3

1. Model of the first Edison talking machine. 2. Edison listening to his battery-operated machine in 1888. 3. The 1888 phonograph used hearing tubes. 4. Emile Berliner. 5. Experimental Phonautograph used by Berliner in 1887. 6. Berliner's 1888 disc-type recording machine. 7. Hand-cranked, disc-type Berliner gramophone of 1893. The metal ball provides connections for hearing tubes. 8. 1903 Columbia disc gramophone.

4

5

7

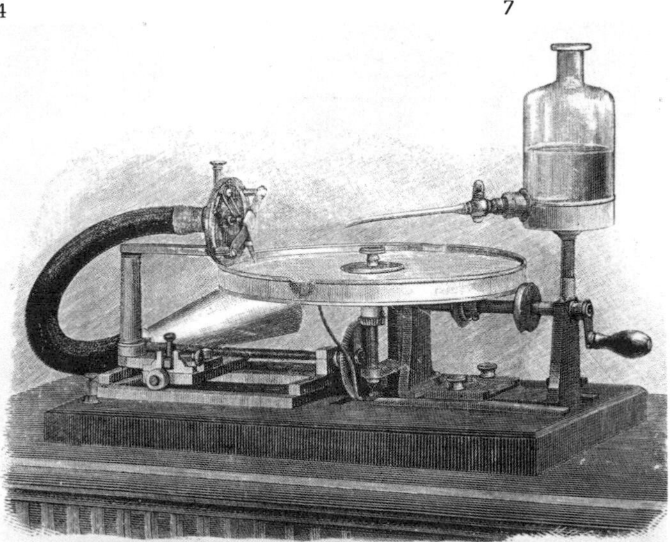

6

8

Right: 1913 Edison "Concert" phonograph. This model, first called the "Opera," was designed to play four-minute cylinders. Although cylinder machines were on their last legs by 1913—Edison introduced a disc machine that same year—owners of old Edison machines could buy new cylinder records until 1929.

Mr. Edison was quick to adopt a wax cylinder—except that Edison's was not cardboard, but a thick-walled cylinder of solid wax. Both instruments were used with hearing tubes. The Bell-Tainter Graphophone was driven by a treadle, like an old-time sewing machine, while Edison's boasted an electric motor. (The spring motor came later, in 1895.) In 1888, however, when Edison's new machine appeared, there were very few places where you could find an electric wall socket to plug into. Batteries were the imperfect answer. Happily for lawyers, the similarities between the machines provided them with litigation for many years.

Emile Berliner, an immigrant from Germany who had invented the improved microphone for Bell's telephone, made the disc phonograph possible also in 1877. Berliner was influenced by two devices—first, an instrument called the Phonautograph invented twenty years earlier by a Frenchman, Léon Scott de Martinville, and secondly by Cros' unbuilt device.

De Martinville's instrument was not meant to reproduce sound audibly. It was a scientific device to show the wave forms of sound. By means of a horn, a membrane, and a stylus, the vibrations caused by sounds were marked on smoke-blackened paper wrapped around a revolving cylinder. Berliner built a similar contraption and then unwrapped the smoke-blackened paper with its sound-drawn markings and had it photoengraved on a thin metal plate, as Cros had proposed. Next he wrapped the engraved plate around the cylinder. When a stylus or needle attached to a diaphragm followed the engraved lines, sound was reproduced.

The wrapping and unwrapping around the cylinder was a nuisance, and, like Cros, Berliner decided to use a transparent disc. He applied a thin layer of lampblack over linseed oil on the bottom surface of a glass disc. He put his lampblack mixture on the lower surface, so that it would drop off where the needle scratched it. This process worked, but it still required photoengraving. In 1888 he eliminated the photographic step by thinly covering a zinc disc with beeswax. The needle cut through the wax, the exposed metal was etched with acid, and the etched plate formed a master from which a number of records could be pressed.

Now there were three kinds of talking machine. But after years of throat-cutting and legal skullduggery, two big companies emerged. The Bell and Berliner patents were the basis of the Columbia Gramophone Company and Edison's company became the Victor Talking Machine Company.

Today Columbia is CBS and Victor is NBC.

8 Epilogue

It has been a longish time since an audacious and beneficial invention has shaken the world. Inventors abound. The pace of invention is unwearied. A vast number of the manufactured products in general use today were unknown before World War II.

But nothing on the scale of the steam engine, the automobile, or the telephone has burst upon us recently to turn our way of living upside down.

The reasons for this dearth of marvels would seem to lie in the increasingly complex and interdependent organization of society. The wonders remaining to be invented may be infinite and beyond imagining. But increasingly they seem to require more brainpower than the individual inventor can muster, and more sophisticated apparatus—computers, lasers, betatrons—than he can afford or operate. And almost the only way his results can be introduced into the marketplace is through the predominant industrial system.

James Watt and Alexander Graham Bell and Fox Talbot and the Wright brothers dreamed their own dreams, and worried, sweated, and pieced together their revolutionary concepts in their own shops, on their own time, and finally, triumphantly, brought their inventions to fruition. Working this way is nearly impossible today. Important inventions in our time cannot be anything so simple as the telephone or the Wright Flyer. Contemporary invention is infinitely more complex, based on abstruse calculations and research far beyond the competence and facilities of the old-time attic experimenter. And increasingly it is only the large corporation that can provide the facilities for research and development of new ideas, and afford the enormous costs involved in product manufacturing, promotion, and distribution.

But corporations prefer a profitable status quo to disruptive change, whatever its promises for mankind's future. As long as the automobiles we are accustomed to continue to sell, manufacturers' interest in a replacement for the internal-combustion engine will be minimal.

Further, many inventions run counter to corporate logic. The razor blade forever sharp, the tire that lasts as long as the car, the endlessly reusable container, the nonobsolescing appliance, the lifetime light bulb . . . One to a customer and the company would be out of business.

Even if an inventor devised something never before seen on earth, it very likely would compete with an existing vested interest. A printing surface other than paper, for instance, might not only alarm the forest-product and paper people, but might raise hob with the entire technology of printing presses, simply because the machinery would not accept the new surface.

Whatever is still to come, there is no doubt that many gaps in our lives have been filled, and that those remaining have been narrowed by the jumble of products consumer-oriented economies dangle before us.

This means that most inventors operating under the shelter of a corporate facility are unlikely to be free-roving spirits whose every brainstorm is welcome, wherever it may lead. Refinements, applications, technological improvements, miniaturization, cost-cutting—these are the routes the usual R & D team is encouraged to run.

The major inventions that have appeared during the lifetime of anyone over forty-five make the point. Frequency-modulation broadcasting, phototypesetting, the electronic computer, the transistor, the Wankel engine, television, instant-picture photography—all, however ingenious, are evolutionary products or processes, merely a step beyond what has been already available.

If this seems a reason for pessimism, it is not. There will always be some brave fellows who will tilt against the formidable resistance of Things As They Are. They will not necessarily produce Great Inventions, but the world needs small inventions, too—things like safety pins and matchbooks and zippers.

And who knows when some kid—another Philo T. Farnsworth—will conjure up a practical, dirt-cheap way to turn sunlight directly into unlimited electric power. Or . . . .

Index

Page numbers in **boldface** refer to illustrations.